CAUSAL INFERENCE IN STATISTICS

CAUSAL INFERENCE IN STATISTICS

A PRIMER

Judea Pearl

Computer Science and Statistics, University of California, Los Angeles, USA

Madelyn Glymour

Philosophy, Carnegie Mellon University, Pittsburgh, USA

Nicholas P. Jewell

Biostatistics and Statistics, University of California, Berkeley, USA

This edition first published 2016

Registered office
John Wiley & Sons Ltd, The Atrium, Southern Gate, Chichester, West Sussex, PO19 8SQ, United Kingdom

For details of our global editorial offices, for customer services and for information about how to apply for permission to reuse the copyright material in this book please see our website at www.wiley.com.

Library of Congress Cataloging-in-Publication Data applied for

ISBN: 9781119186847

A catalogue record for this book is available from the British Library.

Cover Image: © gmaydos/Getty

Typeset in 10/12pt TimesLTStd by SPi Global, Chennai, India

Reprinted with revisions 2021

SKY10034921_062322

To my wife, Ruth, my greatest mentor.
– Judea Pearl

To my parents, who are the causes of me.
– Madelyn Glymour

To Debra and Britta, who inspire me every day.
– Nicholas P. Jewell

Contents

About the Authors

Judea Pearl is Professor of Computer Science and Statistics at the University of California, Los Angeles, where he directs the Cognitive Systems Laboratory and conducts research in artificial intelligence, causal inference and philosophy of science. He is a Co-Founder and Editor of the *Journal of Causal Inference* and the author of three landmark books in inference-related areas.

His groundbreaking book, *Causality: Models, Reasoning and Inference* (Cambridge, 2000, 2009), has introduced many of the methods used in modern causal analysis. It won the Lakatos Award from the London School of Economics and is cited by nearly 20,000 scientific publications. His latest book, *The Book of Why: The New Science of Cause and Effect* (with Dana Mackenzie, Basic Books, 2018), describes the impacts of causal inference to the general public.

Pearl is a member of the National Academy of Sciences, the National Academy of Engineering, a Founding Fellow of the Association for Artificial Intelligence, and Honorary Fellow of the Royal Statistical Society. He is a recipient of numerous prizes and awards, including the Technion's Harvey Prize and the ACM Alan Turing Award for fundamental contributions to probabilistic and causal reasoning.

Madelyn Glymour is a data analyst at Carnegie Mellon University, and a science writer and editor for the Cognitive Systems Laboratory at UCLA. Her interests lie in causal discovery and in the art of making complex concepts accessible to broad audiences.

Nicholas P. Jewell is Professor of Biostatistics and Statistics at the University of California, Berkeley. He has held various academic and administrative positions at Berkeley since his arrival in 1981, most notably serving as Vice Provost from 1994 to 2000. He has also held academic appointments at the University of Edinburgh, Oxford University, the London School of Hygiene and Tropical Medicine, and at the University of Kyoto. In 2007, he was a Fellow at the Rockefeller Foundation Bellagio Study Center in Italy.

Jewell is a Fellow of the American Statistical Association, the Institute of Mathematical Statistics, and the American Association for the Advancement of Science (AAAS). He is a past winner of the Snedecor Award and the Marvin Zelen Leadership Award in Statistical Science from Harvard University. He is currently the Editor of the *Journal of the American Statistical Association – Theory & Methods*, and Chair of the Statistics Section of AAAS. His research focuses on the application of statistical methods to infectious and chronic disease epidemiology, the assessment of drug safety, time-to-event analyses, and human rights.

About the Authors

Preface

When attempting to make sense of data, statisticians are invariably motivated by causal questions. For example, "How effective is a given treatment in preventing a disease?"; "Can one estimate obesity-related medical costs?"; "Could government actions have prevented the financial crisis of 2008?"; "Can hiring records prove an employer guilty of sex discrimination?"

The peculiar nature of these questions is that they cannot be answered, or even articulated, in the traditional language of statistics. In fact, only recently has science acquired a mathematical language we can use to express such questions, with accompanying tools to allow us to answer them from data.

The development of these tools has spawned a revolution in the way causality is treated in statistics and in many of its satellite disciplines, especially in the social and biomedical sciences. For example, in the technical program of the 2003 Joint Statistical Meeting in San Francisco, there were only 13 papers presented with the word "cause" or "causal" in their titles; the number of such papers exceeded 100 by the Boston meeting in 2014. These numbers represent a transformative shift of focus in statistics research, accompanied by unprecedented excitement about the new problems and challenges that are opening themselves to statistical analysis. Harvard's political science professor Gary King puts this revolution in historical perspective: "More has been learned about causal inference in the last few decades than the sum total of everything that had been learned about it in all prior recorded history."

Yet this excitement remains barely seen among statistics educators, and is essentially absent from statistics textbooks, especially at the introductory level. The reasons for this disparity is deeply rooted in the tradition of statistical education and in how most statisticians view the role of statistical inference.

In Ronald Fisher's influential manifesto, he pronounced that "the object of statistical methods is the reduction of data" (Fisher 1922). In keeping with that aim, the traditional task of making sense of data, often referred to generically as "inference," became that of finding a parsimonious mathematical description of the joint distribution of a set of variables of interest, or of specific parameters of such a distribution. This general strategy for inference is extremely familiar not just to statistical researchers and data scientists, but to anyone who has taken a basic course in statistics. In fact, many excellent introductory books describe smart and effective ways to extract the maximum amount of information possible from the available data. These books take the novice reader from experimental design to parameter estimation and hypothesis testing in great detail. Yet the aim of these techniques are invariably the description of data, not of the process responsible for the data. Most statistics books do not even have the word "causal" or "causation" in the index.

Yet the fundamental question at the core of a great deal of statistical inference is causal; do changes in one variable cause changes in another, and if so, how much change do they cause? In avoiding these questions, introductory treatments of statistical inference often fail even to discuss whether the parameters that are being estimated are the relevant quantities to assess when interest lies in cause and effects.

The best that most introductory textbooks do is this: First, state the often-quoted aphorism that "association does not imply causation," give a short explanation of confounding and how "lurking variables" can lead to a misinterpretation of an apparent relationship between two variables of interest. Further, the boldest of those texts pose the principal question: "How can a causal link between x and y be established?" and answer it with the long-standing "gold standard" approach of resorting to randomized experiment, an approach that to this day remains the cornerstone of the drug approval process in the United States and elsewhere.

However, given that most causal questions cannot be addressed through random experimentation, students and instructors are left to wonder if there is anything that can be said with any reasonable confidence in the absence of pure randomness.

In short, by avoiding discussion of causal models and causal parameters, introductory textbooks provide readers with no basis for understanding how statistical techniques address scientific questions of causality.

It is the intent of this primer to fill this gnawing gap and to assist teachers and students of elementary statistics in tackling the causal questions that surround almost any nonexperimental study in the natural and social sciences. We focus here on simple and natural methods to define *causal* parameters that we wish to understand and to show what assumptions are necessary for us to estimate these parameters in observational studies. We also show that these assumptions can be expressed mathematically and transparently and that simple mathematical machinery is available for translating these assumptions into estimable causal quantities, such as the effects of treatments and policy interventions, to identify their testable implications.

Our goal stops there for the moment; we do not address in any detail the optimal parameter estimation procedures that use the data to produce effective statistical estimates and their associated levels of uncertainty. However, those ideas—some of which are relatively advanced—are covered extensively in the growing literature on causal inference. We thus hope that this short text can be used in conjunction with standard introductory statistics textbooks like the ones we have described to show how statistical models and inference can easily go hand in hand with a thorough understanding of causation.

It is our strong belief that if one wants to move beyond mere description, statistical inference cannot be effectively carried out without thinking carefully about causal questions, and without leveraging the simple yet powerful tools that modern analysis has developed to answer such questions. It is also our experience that thinking causally leads to a much more exciting and satisfying approach to both the simplest and most complex statistical data analyses. This is not a new observation. Virgil said it much more succinctly than we in 29 BC:

"Felix, qui potuit rerum cognoscere causas" (Virgil 29 BC)
(Lucky is he who has been able to understand the causes of things)

The book is organized in four chapters.

Chapter 1 provides the basic statistical, probabilistic, and graphical concepts that readers will need to understand the rest of the book. It also introduces the fundamental concepts of causality, including the causal model, and explains through examples how the model can convey information that pure data are unable to provide.

Chapter 2 explains how causal models are reflected in data, through patterns of statistical dependencies. It explains how to determine whether a data set complies with a given causal model, and briefly discusses how one might search for models that explain a given data set.

Chapter 3 is concerned with how to make predictions using causal models, with a particular emphasis on predicting the outcome of a policy intervention. Here we introduce techniques of reducing confounding bias using adjustment for covariates, as well as inverse probability weighting. This chapter also covers mediation analysis and contains an in-depth look at how the causal methods discussed thus far work in a linear system. Key to these methods is the fundamental distinction between regression coefficients and structural parameters, and how students should use both to predict causal effects in linear models.

Chapter 4 introduces the concept of counterfactuals—what would have happened, had we chosen differently at a point in the past—and discusses how we can compute them, estimate their probabilities, and what practical questions we can answer using them. This chapter is somewhat advanced, compared to its predecessors, primarily due to the novelty of the notation and the hypothetical nature of the questions asked. However, the fact that we read and compute counterfactuals using the same scientific models that we used in previous chapters should make their analysis an easy journey for students and instructors. Those wishing to understand counterfactuals on a friendly mathematical level should find this chapter a good starting point, and a solid basis for bridging the model-based approach taken in this book with the potential outcome framework that some experimentalists are pursuing in statistics.

Acknowledgments

This book is an outgrowth of a graduate course on causal inference that the first author has been teaching at UCLA in the past 20 years. It owes many of its tools and examples to former members of the Cognitive Systems Laboratory who participated in the development of this material, both as researchers and as teaching assistants. These include Alex Balke, David Chickering, David Galles, Dan Geiger, Moises Goldszmidt, Jin Kim, George Rebane, Ilya Shpitser, Jin Tian, and Thomas Verma.

We are indebted to many colleagues from whom we have learned much about causal problems, their solutions, and how to present them to general audiences. These include Clark and Maria Glymour, for providing patient ears and sound advice on matters of both causation and writing, Felix Elwert and Tyler VanderWeele for insightful comments on an earlier version of the manuscript, and the many visitors and discussants to the UCLA Causality blog who kept the discussion lively, occasionally controversial, but never boring (causality.cs.ucla.edu/blog).

Elias Bareinboim, Bryant Chen, Andrew Forney, Ang Li, Karthika Mohan, reviewed the text for accuracy and transparency. Ang and Andrew also wrote solutions to the study questions, which are available to instructors from the publisher, see <http://bayes.cs.ucla.edu/PRIMER/CIS-Manual-PUBLIC.pdf>.

The manuscript was most diligently typed, processed, illustrated, and proofed by Kaoru Mulvihill at UCLA. Debbie Jupe and Heather Kay at Wiley deserve much credit for recognizing and convincing us that a book of this scope is badly needed in the field, and for encouraging us throughout the production process.

Finally, the National Science Foundation and the Office of Naval Research deserve acknowledgment for faithfully and consistently sponsoring the research that led to these results, with special thanks to Behzad Kamgar-Parsi.

List of Figures

About the Companion Website

This book is accompanied by a companion website:

www.wiley.com/go/Pearl/Causality

1

Preliminaries: Statistical and Causal Models

1.1 Why Study Causation

The answer to the question "why study causation?" is almost as immediate as the answer to "why study statistics." We study causation because we need to make sense of data, to guide actions and policies, and to learn from our success and failures. We need to estimate the effect of smoking on lung cancer, of education on salaries, of carbon emissions on the climate. Most ambitiously, we also need to understand *how* and *why* causes influence their effects, which is not less valuable. For example, knowing whether malaria is transmitted by mosquitoes or "mal-air," as many believed in the past, tells us whether we should pack mosquito nets or breathing masks on our next trip to the swamps.

Less obvious is the answer to the question, "why study causation as a separate topic, distinct from the traditional statistical curriculum?" What can the concept of "causation," considered on its own, tell us about the world that tried-and-true statistical methods can't?

Quite a lot, as it turns out. When approached rigorously, causation is not merely an aspect of statistics; it is an addition to statistics, an enrichment that allows statistics to uncover workings of the world that traditional methods alone cannot. For example, and this might come as a surprise to many, none of the problems mentioned above can be articulated in the standard language of statistics.

To understand the special role of causation in statistics, let's examine one of the most intriguing puzzles in the statistical literature, one that illustrates vividly why the traditional language of statistics must be enriched with new ingredients in order to cope with cause–effect relationships, such as the ones we mentioned above.

1.2 Simpson's Paradox

Named after Edward Simpson (born 1922), the statistician who first popularized it, the paradox refers to the existence of data in which a statistical association that holds for an entire population is reversed in every subpopulation. For instance, we might discover that students who

Causal Inference in Statistics: A Primer, First Edition. Judea Pearl, Madelyn Glymour, and Nicholas P. Jewell.

Companion Website: www.wiley.com/go/Pearl/Causality

smoke get higher grades, on average, than nonsmokers get. But when we take into account the students' age, we might find that, in every age group, smokers get lower grades than nonsmokers get. Then, if we take into account both age and income, we might discover that smokers once again get *higher* grades than nonsmokers of the same age and income. The reversals may continue indefinitely, switching back and forth as we consider more and more attributes. In this context, we want to decide whether smoking causes grade increases and in which direction and by how much, yet it seems hopeless to obtain the answers from the data.

In the classical example used by Simpson (1951), a group of sick patients are given the option to try a new drug. Among those who took the drug, a lower percentage recovered than among those who did not. However, when we partition by gender, we see that *more* men taking the drug recover than do men are not taking the drug, and more women taking the drug recover than do women are not taking the drug! In other words, the drug appears to help men and women, but hurt the general population. It seems nonsensical, or even impossible—which is why, of course, it is considered a paradox. Some people find it hard to believe that numbers could even be combined in such a way. To make it believable, then, consider the following example:

Example 1.2.1 *We record the recovery rates of 700 patients who were given access to the drug. A total of 350 patients chose to take the drug and 350 patients did not. The results of the study are shown in Table 1.1.*

The first row shows the outcome for male patients; the second row shows the outcome for female patients; and the third row shows the outcome for all patients, regardless of gender. In male patients, drug takers had a better recovery rate than those who went without the drug (93% vs 87%). In female patients, again, those who took the drug had a better recovery rate than nontakers (73% vs 69%). However, in the combined population, those who did not take the drug had a better recovery rate than those who did (83% vs 78%).

The data seem to say that if we know the patient's gender—male or female—we can prescribe the drug, but if the gender is unknown we should not! Obviously, that conclusion is ridiculous. If the drug helps men and women, it must help *anyone*; our lack of knowledge of the patient's gender cannot make the drug harmful.

Given the results of this study, then, should a doctor prescribe the drug for a woman? A man? A patient of unknown gender? Or consider a policy maker who is evaluating the drug's overall effectiveness on the population. Should he/she use the recovery rate for the general population? Or should he/she use the recovery rates for the gendered subpopulations?

Table 1.1 Results of a study into a new drug, with gender being taken into account

	Drug	No drug
Men	81 out of 87 recovered (93%)	234 out of 270 recovered (87%)
Women	192 out of 263 recovered (73%)	55 out of 80 recovered (69%)
Combined data	273 out of 350 recovered (78%)	289 out of 350 recovered (83%)

The answer is nowhere to be found in simple statistics. In order to decide whether the drug will harm or help a patient, we first have to understand the story behind the data—the causal mechanism that led to, or *generated*, the results we see. For instance, suppose we knew an additional fact: Estrogen has a negative effect on recovery, so women are less likely to recover than men, regardless of the drug. In addition, as we can see from the data, women are significantly *more* likely to take the drug than men are. So, the reason the drug appears to be harmful overall is that, if we select a drug user at random, that person is more likely to be a woman and hence less likely to recover than a random person who does not take the drug. Put differently, being a woman is a common cause of both drug taking and failure to recover. Therefore, to assess the effectiveness, we need to compare subjects of the same gender, thereby ensuring that any difference in recovery rates between those who take the drug and those who do not is not ascribable to estrogen. This means we should consult the segregated data, which shows us unequivocally that the drug is helpful. This matches our intuition, which tells us that the segregated data is "more specific," hence more informative, than the unsegregated data.

With a few tweaks, we can see how the same reversal can occur in a continuous example. Consider a study that measures weekly exercise and cholesterol in various age groups. When we plot exercise on the X-axis and cholesterol on the Y-axis and segregate by age, as in Figure 1.1, we see that there is a general trend downward in each group; the more young people exercise, the lower their cholesterol is, and the same applies for middle-aged people and the elderly. If, however, we use the same scatter plot, but we don't segregate by age (as in Figure 1.2), we see a general trend upward; the more a person exercises, the higher their cholesterol is. To resolve this problem, we once again turn to the story behind the data. If we know that older people, who are more likely to exercise (Figure 1.1), are also more likely to have high cholesterol regardless of exercise, then the reversal is easily explained, and easily resolved. Age is a common cause of both treatment (exercise) and outcome (cholesterol). So we should look at the age-segregated data in order to compare same-age people and thereby eliminate the possibility that the high exercisers in each group we examine are more likely to have high cholesterol due to their age, and not due to exercising.

However, and this might come as a surprise to some readers, segregated data does not always give the correct answer. Suppose we looked at the same numbers from our first example of drug taking and recovery, instead of recording participants' gender, patients' blood pressure were

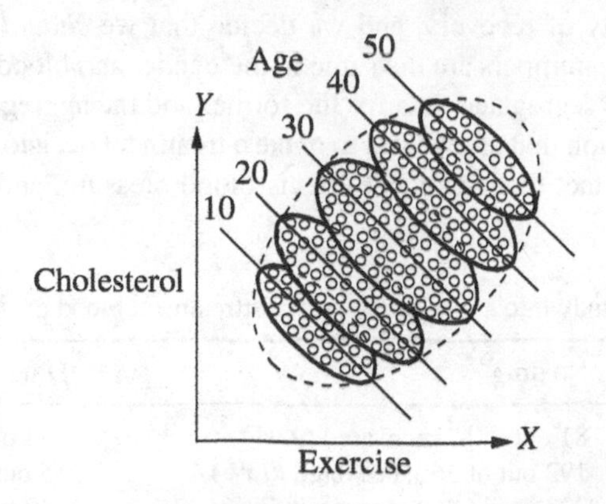

Figure 1.1 Results of the exercise–cholesterol study, segregated by age

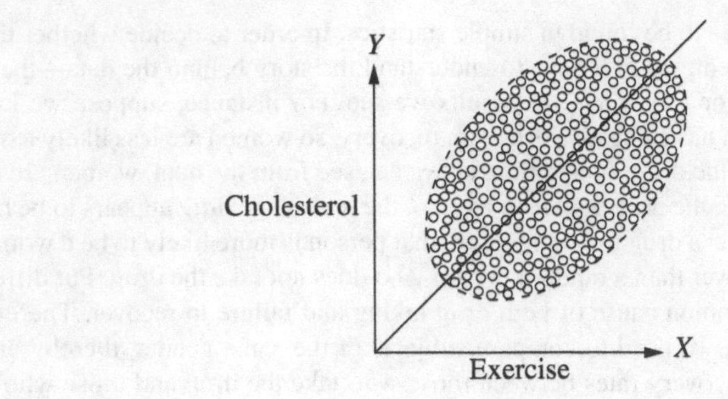

Figure 1.2 Results of the exercise–cholesterol study, unsegregated. The data points are identical to those of Figure 1.1, except the boundaries between the various age groups are not shown

recorded at the end of the experiment. In this case, we know that the drug affects recovery by lowering the blood pressure of those who take it—but unfortunately, it also has a toxic effect. At the end of our experiment, we receive the results shown in Table 1.2. (Table 1.2 is numerically identical to Table 1.1, with the exception of the column labels, which have been switched.)

Now, would you recommend the drug to a patient?

Once again, the answer follows from the way the data were generated. In the general population, the drug might improve recovery rates because of its effect on blood pressure. But in the subpopulations—the group of people whose posttreatment BP is high and the group whose posttreatment BP is low—we, of course, would not see that effect; we would only see the drug's toxic effect.

As in the gender example, the purpose of the experiment was to gauge the overall effect of treatment on rates of recovery. But in this example, since lowering blood pressure is one of the mechanisms by which treatment affects recovery, it makes no sense to separate the results based on blood pressure. (If we had recorded the patients' blood pressure *before* treatment, and if it were BP that had an effect on treatment, rather than the other way around, it would be a different story.) So we consult the results for the general population, we find that treatment increases the probability of recovery, and we decide that we *should* recommend treatment. Remarkably, though the numbers are the same in the gender and blood pressure examples, the correct result lies in the segregated data for the former and the aggregate data for the latter.

None of the information that allowed us to make a treatment decision—not the timing of the measurements, not the fact that treatment affects blood pressure, and not the fact that blood

Table 1.2 Results of a study into a new drug, with posttreatment blood pressure taken into account

	No drug	Drug
Low BP	81 out of 87 recovered (93%)	234 out of 270 recovered (87%)
High BP	192 out of 263 recovered (73%)	55 out of 80 recovered (69%)
Combined data	273 out of 350 recovered (78%)	289 out of 350 recovered (83%)

pressure affects recovery—was found in the data. In fact, as statistics textbooks have traditionally (and correctly) warned students, correlation is not causation, so there is no statistical method that can determine the causal story from the data alone. Consequently, there is no statistical method that can aid in our decision.

Yet statisticians interpret data based on causal assumptions of this kind all the time. In fact, the very paradoxical nature of our initial, qualitative, gender example of Simpson's problem is derived from our strongly held conviction that treatment cannot affect sex. If it could, there would be no paradox, since the causal story behind the data could then easily assume the same structure as in our blood pressure example. Trivial though the assumption "treatment does not cause sex" may seem, there is no way to test it in the data, nor is there any way to represent it in the mathematics of standard statistics. There is, in fact, no way to represent *any* causal information in contingency tables (such as Tables 1.1 and 1.2), on which statistical inference is often based.

There are, however, *extra*-statistical methods that can be used to express and interpret causal assumptions. These methods and their implications are the focus of this book. With the help of these methods, readers will be able to mathematically describe causal scenarios of any complexity, and answer decision problems similar to those posed by Simpson's paradox as swiftly and comfortably as they can solve for X in an algebra problem. These methods will allow us to easily distinguish each of the above three examples and move toward the appropriate statistical analysis and interpretation. A calculus of causation composed of simple logical operations will clarify the intuitions we already have about the nonexistence of a drug that cures men and women but hurts the whole population and about the futility of comparing patients with equal blood pressure. This calculus will allow us to move beyond the toy problems of Simpson's paradox into intricate problems, where intuition can no longer guide the analysis. Simple mathematical tools will be able to answer practical questions of policy evaluation as well as scientific questions of how and why events occur.

But we're not quite ready to pull off such feats of derring-do just yet. In order to rigorously approach our understanding of the causal story behind data, we need four things:

1. A working definition of "causation."
2. A method by which to formally articulate causal assumptions—that is, to create causal models.
3. A method by which to link the structure of a causal model to features of data.
4. A method by which to draw conclusions from the combination of causal assumptions embedded in a model and data.

The first two parts of this book are devoted to providing methods for modeling causal assumptions and linking them to data sets, so that in the third part, we can use those assumptions and data to answer causal questions. But before we can go on, we must define causation. It may seem intuitive or simple, but a commonly agreed-upon, completely encompassing definition of causation has eluded statisticians and philosophers for centuries. For our purposes, the definition of causation is simple, if a little metaphorical: A variable X is a *cause* of a variable Y if Y in any way relies on X for its value. We will expand slightly upon this definition later, but for now, think of causation as a form of listening; X is a cause of Y if Y listens to X and decides its value in response to what it hears.

Readers must also know some elementary concepts from probability, statistics, and graph theory in order to understand the aforementioned causal methods. The next two sections will therefore provide the necessary definitions and examples. Readers with a basic understanding of probability, statistics, and graph theory may skip to Section 1.5 with no loss of understanding.

Study questions

Study question 1.2.1

What is wrong with the following claims?

(a) *"Data show that income and marriage have a high positive correlation. Therefore, your earnings will increase if you get married."*
(b) *"Data show that as the number of fires increase, so does the number of fire fighters. Therefore, to cut down on fires, you should reduce the number of fire fighters."*
(c) *"Data show that people who hurry tend to be late to their meetings. Don't hurry, or you'll be late."*

Study question 1.2.2

A baseball batter Tim has a better batting average than his teammate Frank. However, someone notices that Frank has a better batting average than Tim against both right-handed and left-handed pitchers. How can this happen? (Present your answer in a table.)

Study question 1.2.3

Determine, for each of the following causal stories, whether you should use the aggregate or the segregated data to determine the true effect.

(a) *There are two treatments used on kidney stones: Treatment A and Treatment B. Doctors are more likely to use Treatment A on large (and therefore, more severe) stones and more likely to use Treatment B on small stones. Should a patient who doesn't know the size of his or her stone examine the general population data, or the stone size-specific data when determining which treatment will be more effective?*
(b) *There are two doctors in a small town. Each has performed 100 surgeries in his career, which are of two types: one very difficult surgery and one very easy surgery. The first doctor performs the easy surgery much more often than the difficult surgery and the second doctor performs the difficult surgery more often than the easy surgery. You need surgery, but you do not know whether your case is easy or difficult. Should you consult the success rate of each doctor over all cases, or should you consult their success rates for the easy and difficult cases separately, to maximize the chance of a successful surgery?*

Study question 1.2.4

In an attempt to estimate the effectiveness of a new drug, a randomized experiment is conducted. In all, 50% of the patients are assigned to receive the new drug and 50% to receive a placebo. A day before the actual experiment, a nurse hands out lollipops to some patients who show signs of depression, mostly among those who have been assigned to treatment the next day (i.e., the nurse's round happened to take her through the treatment-bound ward). Strangely, the experimental data revealed a Simpson's reversal: Although the drug proved beneficial to the population as a whole, drug takers were less likely to recover than nontakers, among both lollipop receivers and lollipop nonreceivers. Assuming that lollipop sucking in itself has no effect whatsoever on recovery, answer the following questions:

(a) Is the drug beneficial to the population as a whole or harmful?
(b) Does your answer contradict our gender example, where sex-specific data was deemed more appropriate?
(c) Draw a graph (informally) that more or less captures the story. (Look ahead to Section 1.4 if you wish.)
(d) How would you explain the emergence of Simpson's reversal in this story?
(e) Would your answer change if the lollipops were handed out (by the same criterion) a day after the study?

[Hint: Use the fact that receiving a lollipop indicates a greater likelihood of being assigned to drug treatment, as well as depression, which is a symptom of risk factors that lower the likelihood of recovery.]

1.3 Probability and Statistics

Since statistics generally concerns itself not with absolutes but with likelihoods, the language of probability is extremely important to it. Probability is similarly important to the study of causation because most causal statements are uncertain (e.g., "careless driving causes accidents," which is true, but does not mean that a careless driver is certain to get into an accident), and probability is the way we express uncertainty. In this book, we will use the language and laws of probability to express our beliefs and uncertainty about the world. To aid readers without a strong background in probability, we provide here a glossary of the most important terms and concepts they will need to know in order to understand the rest of the book.

1.3.1 Variables

A *variable* is any property or descriptor that can take multiple values. In a study that compares the health of smokers and nonsmokers, for instance, some variables might be the age of the participant, the gender of the participant, whether or not the participant has a family history of cancer, and how many years the participant has been smoking. A variable can be thought of as a question, to which the value is the answer. For instance, "How old is this participant?" "38 years old." Here, "age" is the variable, and "38" is its value. The probability that variable X takes value x is written $P(X = x)$. This is often shortened, when context allows, to $P(x)$. We can also discuss the probability of multiple values at once; for instance, the probability that $X = x$ and $Y = y$ is written $P(X = x, Y = y)$, or $P(x, y)$. Note that $P(X = 38)$ is specifically interpreted as the probability that an individual randomly selected from the population is aged 38.

A variable can be either *discrete* or *continuous*. Discrete variables (sometimes called *categorical* variables) can take one of a finite or countably infinite set of values in any range. A variable describing the state of a standard light switch is discrete, because it has two values: "on" and "off." Continuous variables can take any one of an infinite set of values on a continuous scale (i.e., for any two values, there is some third value that lies between them). For instance, a variable describing in detail a person's weight is continuous, because weight is measured by a real number.

1.3.2 Events

An *event* is any assignment of a value or set of values to a variable or set of variables. "$X = 1$" is an event, as is "$X = 1$ or $X = 2$," as is "$X = 1$ and $Y = 3$," as is "$X = 1$ or $Y = 3$." "The coin flip lands on heads," "the subject is older than 40," and "the patient recovers" are all events. In the first, "outcome of the coin flip" is the variable, and "heads" is the value it takes. In the second, "age of the subject" is the variable and "older than 40" describes a set of values it may take. In the third, "the patient's status" is the variable and "recovery" is the value. This definition of "event" runs counter to our everyday notion, which requires that some change occur. (For instance, we would not, in everyday conversation, refer to a person being a certain age as an event, but we would refer to that person *turning* a year older as such.) Another way of thinking of an event in probability is this: Any declarative statement (a statement that can be true or false) is an event.

Study questions

Study question 1.3.1

Identify the variables and events invoked in the lollipop story of Study question 1.2.4

1.3.3 Conditional Probability

The probability that some event A occurs, given that we know some other event B has occurred, is the *conditional probability of A given B*. The conditional probability that $X = x$, given that $Y = y$, is written $P(X = x|Y = y)$. As with unconditional probabilities, this is often shortened to $P(x|y)$. Often, the probability that we assign to the event "$X = x$" changes drastically, depending on the knowledge "$Y = y$" that we condition on. For instance, the probability that you have the flu right now is fairly low. But, that probability would become much higher if you were to take your temperature and discover that it is 102 °F.

When dealing with probabilities represented by frequencies in a data set, one way to think of conditioning is *filtering* a data set based on the value of one or more variables. For instance, suppose we looked at the ages of U.S. voters in the last presidential election. According to the Census Bureau, we might get the data set shown in Table 1.3.

In Table 1.3, there were 132,948,000 votes cast in total, so we would estimate that the probability that a given voter was younger than the age of 45 is

$$P(\textit{Voter's Age} < 45) = \frac{20{,}539{,}000 + 30{,}756{,}000}{132{,}448{,}000} = \frac{51{,}295{,}000}{132{,}948{,}000} = 0.38$$

Suppose, however, we want to estimate the probability that a voter was younger than the age of 45, given that we *know* he was elder than the age of 29. To find this out, we simply filter

Table 1.3 Age breakdown of voters in 2012 election (all numbers in thousands)

Age group	# of voters
18–29	20,539
30–44	30,756
45–64	52,013
65+	29,641
	132,948

Table 1.4 Age breakdown of voters over the age of 29 in 2012 election (all numbers in thousands)

Age group	# of voters
30–44	30,756
45–64	52,013
65+	29,641
	112,409

the data to form a new set (shown in Table 1.4), using only the cases where voters were older than 29.

In this new data set, there are 112,409,000 total votes, so we would estimate that

$$P(Voter\ Age < 45 | Voter\ Age > 29) = \frac{30{,}756{,}000}{112{,}409{,}000} = 0.27$$

Conditional probabilities such as these play an important role in investigating causal questions, as we often want to compare how the probability (or, equivalently, risk) of an outcome changes under different filtering, or exposure, conditions. For example, how does the probability of developing lung cancer for smokers compare to the analogous probability for nonsmokers?

Study questions

Study question 1.3.2

Consider Table 1.5 showing the relationship between gender and education level in the U.S. adult population.

(a) Estimate P(High School).
(b) Estimate P(High School OR Female).
(c) Estimate P(High School | Female).
(d) Estimate P(Female | High School).

Table 1.5 The proportion of males and females achieving a given education level

Gender	Highest education achieved	Occurrence (in hundreds of thousands)
Male	Never finished high school	112
Male	High school	231
Male	College	595
Male	Graduate school	242
Female	Never finished high school	136
Female	High school	189
Female	College	763
Female	Graduate school	172

1.3.4 Independence

It might happen that the probability of one event remains unaltered with the observation of another. For example, while observing your high temperature increases the probability that you have the flu, observing that your friend Joe is 38 years old does not change the probability at all. In cases such as this, we say that the two events are *independent*. Formally, events A and B are said to be independent if

$$P(A|B) = P(A) \tag{1.1}$$

that is, the knowledge that B has occurred gives us no additional information about the probability of A occurring. If this equality does not hold, then A and B are said to be *dependent*. Dependence and independence are symmetric relations—if A is dependent on B, then B is dependent on A, and if A is independent of B, then B is independent of A. (Formally, if $P(A|B) = P(A)$, then it *must* be the case that $P(B|A) = P(B)$.) This makes intuitive sense; if "smoke" tells us something about "fire," then "fire" must tell us something about "smoke."

Two events A and B are *conditionally independent* given a third event C if

$$P(A|B, C) = P(A|C) \tag{1.2}$$

and $P(B|A, C) = P(B|C)$. For example, the event "smoke detector is on" is dependent on the event "there is a fire nearby." But these two events may become independent conditional on the third event "there is smoke nearby"; smoke detectors respond to the presence of smoke only, not to its cause. When dealing with data sets, or probability tables, A and B are conditionally independent given C if A and B are independent in the new data set created by filtering on C. If A and B are independent in the original unfiltered data set, they are called *marginally independent*.

Variables, like events, can be dependent or independent of each other. Two variables X and Y are considered independent if for every value x and y that X and Y can take, we have

$$P(X = x|Y = y) = P(X = x) \tag{1.3}$$

(As with independence of events, independence of variables is a symmetrical relation, so it follows that Eq. (1.3) implies $P(Y = y|X = x) = P(Y = y)$.) If for any pair of values of X and Y, this equality does not hold, then X and Y are said to be *dependent*. In this sense, independence of variables can be understood as a set of independencies of events. For instance, "height" and "musical talent" are independent variables; for every height h and level of musical talent m, the probability that a person is h feet high would not change upon discovering that he/she has m amount of talent.

1.3.5 Probability Distributions

A *probability distribution* for a variable X is the set of probabilities assigned to each possible value of X. For instance, if X can take three values—1, 2, and 3—a possible probability distribution for X would be "$P(X = 1) = 0.5, P(X = 2) = 0.25, P(X = 3) = 0.25$." The probabilities in a probability distribution must lie between 0 and 1, and must sum to 1. An event with probability 0 is impossible; an event with probability 1 is certain.

Continuous variables also have probability distributions. The probability distribution of a continuous variable X is represented by a function f, called the *density function*. When f is plotted on a coordinate plane, the probability that the value of variable X lies between values a and b is the area under the curve between a and b—or, as those who have taken calculus will know, $\int_a^b f(x)dx$. The area under the entire curve—that is, $\int_{-\infty}^{\infty} f(x)dx$—must of course be equal to 1.

Sets of variables can also have probability distributions, called *joint distributions*. The joint distribution of a set of variables V is the set of probabilities of each possible combination of variable values in V. For instance, if V is a set of two variables—X and Y—each of which can take two values—1 and 2—then one possible joint distribution for V is "$P(X = 1, Y = 1) = 0.2, P(X = 1, Y = 2) = 0.1, P(X = 2, Y = 1) = 0.5, P(X = 2, Y = 2) = 0.2$." Just as with single-variable distributions, probabilities in a joint distribution must sum to 1.

1.3.6 The Law of Total Probability

There are several universal probabilistic truths that are useful to know. First, for any two mutually exclusive events A and B (i.e., A and B cannot co-occur), we have

$$P(A \text{ or } B) = P(A) + P(B) \tag{1.4}$$

It follows that, for any two events A and B, we have

$$P(A) = P(A, B) + P(A, \text{"not } B\text{"}) \tag{1.5}$$

because the events "A and B" and "A and 'not B'" are mutually exclusive—and because if A is true, then either "A and B" or "A and 'not B'" must be true. For example, "Dana is a tall man" and "Dana is a tall woman" are mutually exclusive, and if Dana is tall, then he or she must be either a tall man or a tall woman; therefore, $P(\text{"Dana is tall"}) = P(\text{"Dana is a tall man"}) + P(\text{"Dana is a tall woman"})$.

More generally, for any set of events $B_1, B_2, \ldots, B_n$ such that exactly one of the events must be true (an exhaustive, mutually exclusive set, called a *partition*), we have

$$P(A) = P(A, B_1) + P(A, B_2) + \cdots + P(A, B_n) \tag{1.6}$$

This rule, known as the *law of total probability*, becomes somewhat obvious as soon as we put it in real-world terms: If we pull a random card from a standard deck, the probability that the card is a Jack will be equal to the probability that it's a Jack *and* a spade, plus the probability that it's a Jack *and* a heart, plus the probability that it's a Jack *and* a club, plus the probability that it's a Jack *and* a diamond. Calculating the probability of an event A by summing up its probabilities over all B_i is called *marginalizing over B*, and the resulting probability $P(A)$ is called the *marginal probability* of A.

If we know the probability of B and the probability of A conditional on B, we can deduce the probability of A and B by simple multiplication:

$$P(A, B) = P(A|B)P(B) \tag{1.7}$$

For instance, the probability that Joe is funny and smart is equal to the probability that a smart person is funny, multiplied by the probability that Joe is smart. The division rule

$$P(A|B) = P(A, B)/P(B)$$

which is formally regarded as a definition of conditional probabilities, is justified by viewing conditioning as a filtering operation, as we have done in Tables 1.3 and 1.4. When we condition on B, we remove from the table all events that conflict with B. The resulting subtable, like the original, represents a probability distribution, and like all probability distributions, it must sum to one. Since the probabilities of the subtables rows in the original distribution summed to $P(B)$ (by definition), we can determine their probabilities in the new distribution by multiplying each by $1/P(B)$.

Equation (1.7) implies that the notion of independence, which until now we have used informally to mean "giving no additional information," has a numerical representation in the probability distribution. In particular, for events A and B to be independent, we require that

$$P(A, B) = P(A)P(B)$$

For example, to check if the outcomes of two coins are truly independent, we should count the frequency at which both show up tails, and make sure that it equals the product of the frequencies at which each of the coins shows up tails.

Using (1.7) together with the symmetry $P(A, B) = P(B, A)$, we can immediately obtain one of the most important laws of probability, *Bayes' rule*:

$$P(A|B) = \frac{P(B|A)P(A)}{P(B)} \tag{1.8}$$

With the help of the multiplication rule in (1.7), we can express the law of total probability as a weighted sum of conditional probabilities:

$$P(A) = P(A|B_1)P(B_1) + P(A|B_2)P(B_2) + \cdots + P(A|B_k)P(B_k) \tag{1.9}$$

This is very useful, because often we will find ourselves in a situation where we cannot assess $P(A)$ directly, but we can through this decomposition. It is generally easier to assess conditional probabilities such as $P(A|B_k)$, which are tied to specific contexts, rather than $P(A)$, which is not attached to a context. For instance, suppose we have a stock of gadgets from two sources: 30% of them are manufactured by factory A, in which one out of 5000 is defective, whereas 70%

are manufactured by factory B, in which one out of 10,000 is defective. To find the probability that a randomly chosen gadget will be defective is not a trivial mental task, but when broken down according to Eq. (1.9) it becomes easy:

$$\begin{aligned}
P(defective) &= P(defective|A)P(A) + P(defective|B)P(B) \\
&= \frac{0.30}{5{,}000} + \frac{0.70}{10{,}000} \\
&= \frac{1.30}{10{,}000} = 0.00013
\end{aligned}$$

Or, to take a somewhat harder example, suppose we roll two dice, and we want to know the probability that the second roll is higher than the first, $P(A) = P(Roll\ 2 > Roll\ 1)$. There is no obvious way to calculate this probability all at once. But if we break it down into contexts $B_1, \ldots, B_6$ by conditioning on the value of the first die, it becomes easy to solve:

$$\begin{aligned}
P(Roll\ 2 > Roll\ 1) &= P(Roll\ 2 > Roll\ 1|Roll\ 1 = 1)P(Roll\ 1 = 1) \\
&\quad + P(Roll\ 2 > Roll\ 1|Roll\ 1 = 2)P(Roll\ 1 = 2) \\
&\quad + \cdots + P(Roll\ 2 > Roll\ 1|Roll\ 1 = 6) \times P(Roll\ 1 = 6) \\
&= \left(\frac{5}{6} \times \frac{1}{6}\right) + \left(\frac{4}{6} \times \frac{1}{6}\right) + \left(\frac{3}{6} \times \frac{1}{6}\right) + \left(\frac{2}{6} \times \frac{1}{6}\right) + \left(\frac{1}{6} \times \frac{1}{6}\right) + \left(\frac{0}{6} \times \frac{1}{6}\right) \\
&= \frac{5}{12}
\end{aligned}$$

The decomposition described in Eq. (1.9) is sometimes called "the law of alternatives" or "extending the conversation"; in this book, we will refer to it as *conditionalizing on B*.

1.3.7 Using Bayes' Rule

When using Bayes' rule, we sometimes loosely refer to event A as the "hypothesis" and event B as the "evidence." This naming reflects the reason that Bayes' theorem is so important: In many cases, we know or can easily determine $P(B|A)$ (the probability that a piece of evidence will occur, given that our hypothesis is correct), but it's much harder to figure out $P(A|B)$ (the probability of the hypothesis being correct, given that we obtain a piece of evidence). Yet the latter is the question that we most often want to answer in the real world; generally, we want to update our belief in some hypothesis, $P(A)$, after some evidence B has occurred, to $P(A|B)$. To precisely use Bayes' rule in this manner, we must treat each hypothesis as an event and assign to all hypotheses for a given situation a probability distribution, called a *prior*.

For example, suppose you are in a casino, and you hear a dealer shout "11!" You happen to know that the only two games played at the casino that would occasion that event are craps and roulette and that there are exactly as many craps games as roulette games going on at any moment. What is the probability that the dealer is working at a game of craps, given that he shouted "11?"

In this case, "craps" is our hypothesis, and "11" is our evidence. It's difficult to figure out this probability off-hand. But the reverse—the probability that an 11 will result in a given round of craps—is easy to calculate; it is specified by the game. Craps is a game in which gamblers bet on the sum of a roll of two dice. So 11 will be the sum in $\frac{2}{36} = \frac{1}{18}$ of cases: $P(\text{"11"}|\text{"}craps\text{"}) = \frac{1}{18}$. In roulette, there are 38 equally probable outcomes, so $P(\text{"11"}|\text{"}roulette\text{"}) = \frac{1}{38}$. In this situation, there are two possible hypotheses; "craps" and "roulette." Since there are an equal number of craps and roulette games, $P(\text{"}craps\text{"}) = \frac{1}{2}$, our prior belief before we hear the "11" shout. Using the law of total probability,

$$P(\text{"11"}) = P(\text{"11"}|\text{"}craps\text{"})P(\text{"}craps\text{"}) + P(\text{"11"}|\text{"}roulette\text{"})P(\text{"}roulette\text{"})$$

$$= \frac{1}{2} \times \frac{1}{18} + \frac{1}{2} \times \frac{1}{38} = \frac{7}{171}$$

We have now fairly easily obtained all the information we need to determine $P(\text{"}craps\text{"}|\text{"11"})$:

$$P(\text{"}craps\text{"}|\text{"11"}) = \frac{P(\text{"11"}|\text{"}craps\text{"}) \times P(\text{"}craps\text{"})}{P(\text{"11"})} = \frac{1/18 \times 1/2}{7/171} = 0.679$$

Another informative example of Bayes' rule in action is the Monty Hall problem, a classic brain teaser in statistics. In the problem, you are a contestant on a game show, hosted by Monty Hall. Monty shows you three doors—A, B, and C—behind one and only one of which is a new car. (The other two doors have goats.) If you guess correctly, the car is yours; otherwise, you get a goat. You guess A at random. Monty, who is forbidden from revealing where the car is, then opens Door C, which, of course, has a goat behind it. He tells you that you can now switch to Door B, or stick with Door A. Whichever you pick, you'll get what's behind it.

Are you better off opening Door A, or switching to Door B?

Many people, when they first encounter the problem, reason that, since the location of the car is independent of the door you first choose, switching doors neither gains nor loses you anything; the probability that the car is behind Door A is equal to the probability that it is behind Door B.

But the correct answer, as decades of statistics students have found to their consternation, is that you are twice as likely to win the car if you switch to Door B as you are if you stay with Door A. The reasoning often given for this counterintuitive solution is that, when you originally chose a door, you had a $\frac{1}{3}$ probability of picking the door with the car. Since Monty *always* opens a door with a goat, no matter whether you initially chose the car or not, you have received no new information since then. Therefore, there is still a $\frac{1}{3}$ probability that the door you picked hides the car, and the remaining $\frac{2}{3}$ probability must lie with the only other closed door left.

We can prove this surprising fact using Bayes' rule. Here we have three variables: X, the door chosen by the player; Y, the door behind which the car is hidden; and Z, the door which the host opens. X, Y, and Z can all take the values A, B, or C. We want to prove that $P(Y = B|X = A, Z = C) > P(Y = A|X = A, Z = C)$. Our hypothesis is that the car lies behind Door A; our evidence is that Monty opened Door C. We will leave the proof to the reader—see Study question 1.3.5. To further develop your intuition, you might generalize the game to having 100 doors (which contain 1 hidden car and 99 hidden goats). The contestant still chooses one door, but now Monty opens 98 doors—all revealing goats deliberately—before offering the contestant the chance to switch before the final doors are opened. Now, the choice to switch should be obvious.

Why does Monty opening Door C constitute evidence about the location of the car? It didn't, after all, provide any evidence for whether your initial choice of door was correct. And, surely, when he was about to open a door, be it *B* or *C*, you knew in advance that you won't find a car behind it. The answer is that there was no way for Monty to open Door *A* after you chose it—but he *could* have opened Door *B*. The fact that he didn't makes it more likely that he opened Door *C* because he was forced to; it provides evidence that the car lies behind Door *B*. This is a general theme of Bayesian analysis: Any hypothesis that has withstood some test of refutation becomes more likely. Door *B* was vulnerable to refutation (i.e., Monty could have opened it), but Door *A* was not. Therefore, Door *B* becomes a more likely location, whereas Door *A* does not.

The reader may find it instructive to note that the explanation above is laden with counterfactual terminology; for example, "He could have opened," "because he was forced," "He was about to open." Indeed, what makes the Monty Hall example unique among probability puzzles is its critical dependence on the process that generated the data. It shows that our beliefs should depend not merely on the facts observed but also on the process that led to those facts. In particular, the information that the car is not behind Door *C*, in itself, is not sufficient to describe the problem; to figure out the probabilities involved, we must also know what options were available to the host before opening Door *C*. In Chapter 4 of this book we will formulate a theory of counterfactuals that will enable us to describe such processes and alternative options, so as to form the correct beliefs about choices.

There is some controversy attached to Bayes' rule. Often, when we are trying to ascertain the probability of a hypothesis given some evidence, we have no way to calculate the prior probability of the hypothesis, $P(A)$, in terms of fractions or frequencies of cases. Consider: If we did not know the proportion of roulette tables to craps tables in the casino, how on Earth could we determine the prior probability *P*("*craps*")? We might be tempted to postulate $P(A) = \frac{1}{2}$ as a way of expressing our ignorance. But what if we have a hunch that roulette tables are less common in this casino, or the tone of the voice of the caller reminds us of a craps dealer we heard yesterday? In cases such as this, in order to use Bayes' rule, we substitute, in place of $P(A)$, our *subjective belief* in the relative truth of the hypothesis compared to other possibilities. The controversy stems from the subjective nature of that belief—how are we to know whether the assigned $P(A)$ accurately summarizes the information we have about the hypothesis? Should we insist on distilling all of our pro and con arguments down to a single number? And even if we do, why should we update our subjective beliefs about hypotheses the same way that we update objective frequencies? Some behavioral experiments suggest that people do not update their beliefs in accordance with Bayes' rule—but many believe that they *should*, and that deviations from the rule represent compromises, if not deficiencies in reasoning, and lead to suboptimal decisions. Debate over the proper use of Bayes' theorem continues to this day. Despite these controversies, however, Bayes' rule is a powerful tool for statistics, and we will use it to great effect throughout this book.

Study questions

Study question 1.3.3

Consider the casino problem described in Section 1.3.6

(a) *Compute P(*"craps"*|*"*11*"*) assuming that there are twice as many roulette tables as craps games at the casino.*

(b) Compute P("roulette"|"10") assuming that there are twice as many craps games as roulette tables at the casino.

Study question 1.3.4

Suppose we have three cards. Card 1 has two black faces, one on each side; Card 2 has two white faces; and Card 3 has one white face and one black face. You select a card at random and place it on the table. You find that it is black on the face-up side. What is the probability that the face-down side of the card is also black?

(a) Use your intuition to argue that the probability that the face-down side of the card is also black is $\frac{1}{2}$. Why might it be greater than $\frac{1}{2}$?
(b) Express the probabilities and conditional probabilities that you find easy to estimate (for example, $P(C_D = Black)$), in terms of the following variables:

$$I = \text{Identity of the card selected (Card 1, Card 2, or Card 3)}$$

$$C_D = \text{Color of the face-down side (Black, White)}$$

$$C_U = \text{Color of the face-up side (Black, White)}$$

Find the probability that the face-down side of the selected card is black, using your estimates above.
(c) Use Bayes' theorem to find the correct probability of a randomly selected card's back being black if you observe that its front is black?

Study question 1.3.5 (Monty Hall)

Prove, using Bayes' theorem, that switching doors improves your chances of winning the car in the Monty Hall problem.

1.3.8 Expected Values

In statistics, one often deals with data sets and probability distributions that are too large to effectively examine each possible combination of values. Instead, we use statistical measures to represent, with some loss of information, meaningful features of the distribution. One such measure is the *expected value*, also called the *mean*, which can be used when variables take on numerical values. The expected value of a variable X, denoted $E(X)$, is found by multiplying each possible value of the variable by the probability that the variable will take that value, then summing the products:

$$E(X) = \sum_x x\,P(X = x) \tag{1.10}$$

For instance, a variable X representing the outcome of one roll of a fair six-sided die has the following probability distribution: $P(1) = \frac{1}{6}, P(2) = \frac{1}{6}, P(3) = \frac{1}{6}, P(4) = \frac{1}{6}, P(5) = \frac{1}{6}, P(6) = \frac{1}{6}$. The expected value of X is given by:

$$E(X) = \left(1 \times \frac{1}{6}\right) + \left(2 \times \frac{1}{6}\right) + \left(3 \times \frac{1}{6}\right) + \left(4 \times \frac{1}{6}\right) + \left(5 \times \frac{1}{6}\right) + \left(6 \times \frac{1}{6}\right) = 3.5$$

Similarly, the expected value of any function of X—say, $g(X)$—is obtained by summing $g(x)P(X = x)$ over all values of X.

$$E[g(X)] = \sum_x g(x)P(x) \tag{1.11}$$

For example, if after rolling a die, I receive a cash prize equal to the square of the result, we have $g(X) = X^2$, and the expected prize is

$$E[g(X)] = \left(1^2 \times \frac{1}{6}\right) + \left(2^2 \times \frac{1}{6}\right) + \left(3^2 \times \frac{1}{6}\right) + \left(4^2 \times \frac{1}{6}\right) + \left(5^2 \times \frac{1}{6}\right) + \left(6^2 \times \frac{1}{6}\right) = 15.17 \tag{1.12}$$

We can also calculate the expected value of Y conditional on X, $E(Y|X = x)$, by multiplying each possible value y of Y by $P(Y = y|X = x)$, and summing the products.

$$E(Y|X = x) = \sum_y y\, P(Y = y|X = x) \tag{1.13}$$

$E(X)$ is one way to make a "best guess" of X's value. Specifically, out of all the guesses g that we can make, the choice "$g = E(X)$" minimizes the expected square error $E(g - X)^2$. Similarly, $E(Y|X = x)$ represents a best guess of Y, given that we observe $X = x$. If $g = E(Y|X = x)$, then g minimizes the expected square error $E[(g - Y)^2|X = x]$.

For example, the expected age of a 2012 voter, as demonstrated by Table 1.3, is

$$E(\textit{Voter's Age}) = 23.5 \times 0.16 + 37 \times 0.23 + 54.5 \times 0.39 + 70 \times 0.22 = 48.9$$

(For this calculation, we have assumed that every age within each category is equally likely, e.g., a voter is as likely to be 18 as 25, and as likely to be 30 as 44. We have also assumed that the oldest age of any voter is 75.) This means that if we were asked to guess the age of a randomly chosen voter, with the understanding that if we were off by e years, we would lose e^2 dollars, we would lose the least money, on average, if we guessed 48.9. Similarly, if we were asked to guess the age of a random voter younger than the age of 45, our best bet would be

$$E[\textit{Voter's Age} \mid \textit{Voter's Age} < 45] = 23.5 \times 0.40 + 37 \times 0.60 = 31.6 \tag{1.14}$$

The use of expectations as a basis for predictions or "best guesses" hinges to a great extent on an implicit assumption regarding the distribution of X or $Y|X = x$, namely that such distributions are approximately *symmetric*. If, however, the distribution of interest is highly *skewed*, other methods of prediction may be better. In such cases, for example, we might use the median of the distribution of X as our "best guess"; this estimate minimizes the expected absolute error $E(|g - X|)$. We will not pursue such alternative measures further here.

1.3.9 Variance and Covariance

The *variance* of a variable X, denoted $Var(X)$ or σ_X^2, is a measure of roughly how "spread out" the values of X in a data set or population are from their mean. If the values of X all hover close to one value, the variance will be relatively small; if they cover a large range, the variance will be comparatively large. Mathematically, we define the variance of a variable as the average square difference of that variable from its mean. It can be computed by first finding its mean, μ, and then calculating

$$Var(X) = E((X - \mu)^2) \quad (1.15)$$

The *standard deviation* σ_X of a random variable X is the square root of its variance. Unlike the variance, σ_X is expressed in the same units as X. For example, the variance of under-45 voters' age distribution, according to Table 1.3, can easily be calculated to be (Eq. (1.15)):

$$\begin{aligned} Var(X) &= ((23.5 - 31.5)^2 \times 0.41) + ((37 - 31.5)^2 \times 0.59) \\ &= (64 \times 0.41) + (30.25 \times .59) \\ &= 26.24 + 17.85 = 43.09 \text{ years}^2 \end{aligned}$$

while the standard deviation is

$$\sigma_X = \sqrt{(43.09)} = 6.56 \text{ years}$$

This means that, choosing a voter at random, chances are high that his/her age will fall less than 6.56 years away from the average 31.5. This kind of interpretation can be quantified. For example, for a normally distributed random variable X, approximately two-thirds of the population values of X fall within *one* standard deviation of the expectation, or mean. Further, about 95% fall within *two* standard deviations from the mean.

Of special importance is the expectation of the product $(X - E(X))(Y - E(Y))$, which is known as the *covariance* of X and Y,

$$\sigma_{XY} \triangleq E[(X - E(X))(Y - E(Y))] \quad (1.16)$$

It measures the degree to which X and Y *covary*, that is, the degree to which the two variables vary together, or are "associated." This measure of association actually reflects a specific way in which X and Y covary; it measures the extent to which X and Y *linearly* covary. You can think of this as plotting Y versus X and considering the extent to which a straight *line* captures the way in which Y varies as X changes.

The covariance σ_{XY} is often normalized to yield the *correlation coefficient*

$$\rho_{XY} = \frac{\sigma_{XY}}{\sigma_X \sigma_Y} \quad (1.17)$$

which is a dimensionless number ranging from -1 to 1, which represents the slope of the best-fit line after we normalize both X and Y by their respective standard deviations. ρ_{XY} is one if and only if one variable can predict the other in a *linear* fashion, and it is zero whenever such a linear prediction is no better than a random guess. The significance of σ_{XY} and ρ_{XY} will be discussed in the next section. At this point, it is sufficient to note that these degrees of covariation can be readily computed from the joint distribution $P(x, y)$, using Eqs. (1.16) and (1.17). Moreover, both σ_{XY} and ρ_{XY} vanish when X and Y are independent. Note that nonlinear relationships between Y and X cannot naturally be captured by a simple numerical summary; they require a full specification of the conditional probability $P(Y = y|X = x)$.

Study questions

Study question 1.3.6

(a) Prove that, in general, both σ_{XY} and ρ_{XY} vanish when X and Y are independent. [Hint: Use Eqs. (1.16) and (1.17).]
(b) Give an example of two variables that are highly dependent and, yet, their correlation coefficient vanishes.

Study question 1.3.7

Two fair coins are flipped simultaneously to determine the payoffs of two players in the town's casino. Player 1 wins a dollar if and only if at least one coin lands on head. Player 2 receives a dollar if and only if the two coins land on the same face. Let X stand for the payoff of Player 1 and Y for the payoff of Player 2.

(a) Find and describe the probability distributions

$$P(x), P(y), P(x,y), P(y|x) \text{ and } P(x|y)$$

(b) Using the descriptions in (a), compute the following measures:

$$E[X], E[Y], E[Y|X=x], E[X|Y=y]$$
$$Var(X), Var(Y), Cov(X,Y), \rho_{XY}$$

(c) Given that Player 2 won a dollar, what is your best guess of Player 1's payoff?
(d) Given that Player 1 won a dollar, what is your best guess of Player 2's payoff?
(e) Are there two events, $X = x$ and $Y = y$, that are mutually independent?

Study question 1.3.8

Compute the following theoretical measures of the outcome of a single game of craps (one roll of two independent dice), where X stands for the outcome of Die 1, Z for the outcome of Die 2, and Y for their sum.

(a)

$$E[X], E[Y], E[Y|X=x], E[X|Y=y], \text{ for each value of x and y, and}$$
$$Var(X), Var(Y), Cov(X,Y), \rho_{XY}, Cov(X,Z)$$

Table 1.6 describes the outcomes of 12 craps games.

(b) Find the sample estimates of the measures computed in (a), based on the data from Table 1.6. [Hint: Many software packages are available for doing this computation for you.]
(c) Use the results in (a) to determine the best estimate of the sum, Y, given that we measured $X = 3$.

Table 1.6 Results of 12 rolls of two fair dice

	X Die 1	Z Die 2	Y Sum
Roll 1	6	3	9
Roll 2	3	4	7
Roll 3	4	6	10
Roll 4	6	2	8
Roll 5	6	4	10
Roll 6	5	3	8
Roll 7	1	5	6
Roll 8	3	5	8
Roll 9	6	5	11
Roll 10	3	5	8
Roll 11	5	3	8
Roll 12	4	5	9

(d) *What is the best estimate of X, given that we measured $Y = 4$?*
(e) *What is the best estimate of X, given that we measured $Y = 4$ and $Z = 1$? Explain why it is not the same as in (d).*

1.3.10 Regression

Often, in statistics, we wish to predict the value of one variable, Y, based on the value of another variable, X. For example, we may want to predict a student's height based on his age. We noted earlier that the best prediction of Y based on X is given by the conditional expectation $E[Y|X = x]$, at least in terms of mean-squared error. But this assumes that we know the conditional expectation, or can compute it, from the joint distribution $P(y, x)$. With *regression*, we make our prediction directly from the data. We try to find a formula, usually a linear function, that takes observed values of X as input and gives values of Y as output, such that the square error between the predicted and actual values of Y is minimized, on average.

We start with a scatter plot that takes every case in our data set and charts them on a coordinate plane, as shown in Figure 1.2. Our predictor, or input, variable goes on the x-axis, and the variable whose value we are predicting goes on the y-axis.

The least squares regression line is the line for which the sum of the squared vertical distances of the points on the scatter plot from the line is minimized. That is, if there are n data points (x, y) on our scatter plot, and for any data point (x_i, y_i), the value y'_i represents the value of the line $y = \alpha + \beta x$ at x_i, then the least squares regression line is the one that minimizes the value

$$\sum_i (y_i - y'_i)^2 = \sum_i (y_i - \alpha - \beta x_i)^2 \tag{1.18}$$

To see how the slope β relates to the probability distribution $P(x, y)$, suppose we play 12 successive rounds of craps, and get the results shown in Table 1.6. If we wanted to predict the sum Y of the die rolls based on the value of $X = Die\ 1$ alone, using the data in Table 1.6, we would use the scatter plot shown in Figure 1.3. For our craps example, the least squares

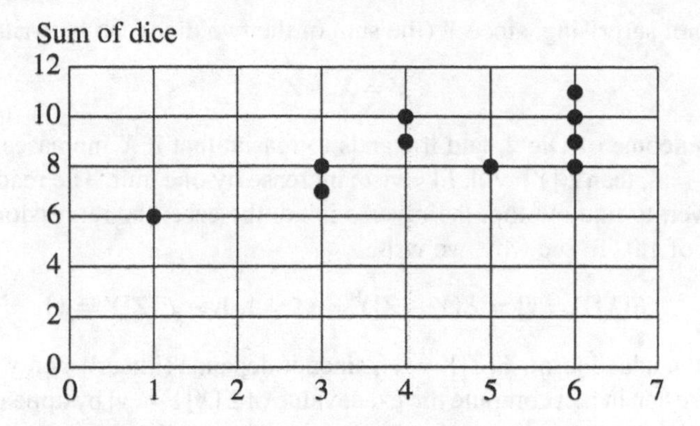

Figure 1.3 Scatter plot of the results in Table 1.6, with the value of Die 1 on the x-axis and the sum of the two dice rolls on the y-axis

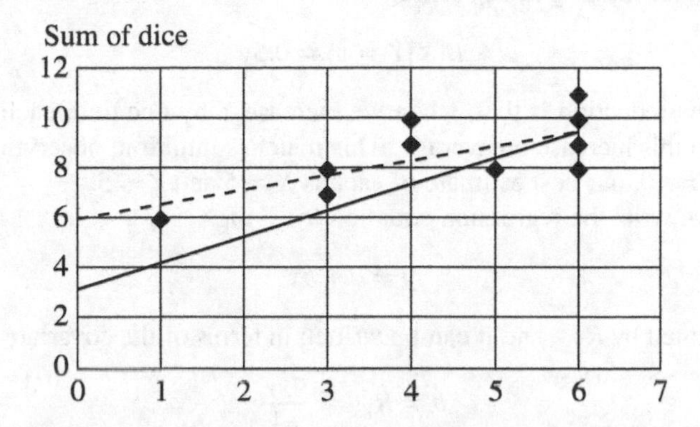

Figure 1.4 Scatter plot of the results in Table 1.6, with the value of Die 1 on the x-axis and the sum of the two dice rolls on the y-axis. The dotted line represents the line of best fit based on the data. The solid line represents the line of best fit we would expect in the population

regression line is shown in Figure 1.4. Note that the regression line for the *sample* that we used is not necessarily the same as the regression line for the *population*. The population is what we get when we allow our sample size to increase to infinity. The solid line in Figure 1.4 represents the theoretical least-square line, which is given by

$$y = 3.5 + 1.0x \tag{1.19}$$

The dashed line represents the sample least-square line, which, due to sampling variations, differs from the theoretical both in slope and in intercept.

In Figure 1.4, we know the equation of the regression line for the population because we know the expected value of the sum of two dice rolls, given that the first die lands on x. The computation is simple:

$$E[Y|X = x] = E[Die\ 2 + X|X = x] = E[Die\ 2] + x = 3.5 + 1.0x$$

This result is not surprising, since Y (the sum of the two dice) can be written as

$$Y = X + Z$$

where Z is the outcome of Die 2, and it stands to reason that if X increases by one unit, say from $X = 3$ to $X = 4$, then $E[Y]$ will, likewise, increase by one unit. The reader might be a bit surprised, however, to find out that the reverse is not the case; the regression of X on Y does not have a slope of 1.0. To see why, we write

$$E[X|Y = y] = E[Y - Z|Y = y] = 1.0y - E[Z|Y = y] \tag{1.20}$$

and realize that the added term, $E[Z|Y = y]$, since it depends (linearly) on y, makes the slope less than unity. We can in fact compute the exact value of $E[X|Y = y]$ by appealing to symmetry and write

$$E[X|Y = y] = E[Z|Y = y]$$

which gives, after substituting in Eq. (1.20),

$$E[X|Y = y] = 0.5y$$

The reason for this reduction is that, when we increase Y by one unit, each of X and Z contributes equally to this increase on average. This matches intuition; observing that the sum of the two dice is $Y = 10$, our best estimate of each is $X = 5$ and $Z = 5$.

In general, if we write the regression equation for Y on X as

$$y = a + bx \tag{1.21}$$

the slope b is denoted by R_{YX}, and it can be written in terms of the covariate σ_{XY} as follows:

$$b = R_{YX} = \frac{\sigma_{XY}}{\sigma_X^2} \tag{1.22}$$

From this equation, we see clearly that the slope of Y on X may differ from the slope of X on Y—that is, in most cases, $R_{YX} \neq R_{XY}$. ($R_{YX} = R_{XY}$ only when the variance of X is equal to the variance of Y.) The slope of the regression line can be positive, negative, or zero. If it is positive, X and Y are said to have a *positive correlation*, meaning that as the value of X gets higher, the value of Y gets higher; if it is negative, X and Y are said to have a *negative correlation*, meaning that as the value of X gets higher, the value of Y gets lower; if it is zero (a horizontal line), X and Y have no linear correlation, and knowing the value of X does not assist us in predicting the value of Y, at least linearly. If two variables are correlated, whether positively or negatively (or in some other way), they are dependent.

1.3.11 Multiple Regression

It is also possible to regress a variable on several variables, using *multiple linear regression*. For instance, if we wanted to predict the value of a variable Y using the values of the variables X and Z, we could perform multiple linear regression of Y on $\{X, Z\}$, and estimate a regression relationship

$$y = r_0 + r_1x + r_2z \tag{1.23}$$

which represents an inclined plane through the three-dimensional coordinate system.

We can create a three-dimensional scatter plot, with values of Y on the y-axis, X on the x-axis, and Z on the z-axis. Then, we can cut the scatter plot into slices along the Z-axis. Each slice will constitute a two-dimensional scatter plot of the kind shown in Figure 1.4. Each of those 2-D scatter plots will have a regression line with a slope r_1. Slicing along the X-axis will give the slope r_2.

The slope of Y on X when we hold Z constant is called the *partial regression coefficient* and is denoted by $R_{YX \cdot Z}$. Note that it is possible for R_{YX} to be positive, whereas $R_{YX \cdot Z}$ is negative as shown in Figure 1.1. This is a manifestation of Simpson's Paradox: positive association between Y and X overall, that becomes negative when we condition on the third variable Z.

The computation of partial regression coefficients (e.g., r_1 and r_2 in (1.23)) is greatly facilitated by a theorem that is one of the most fundamental results in regression analysis. It states that if we write Y as a linear combination of variables $X_1, X_2, \ldots, X_k$ plus a noise term ϵ,

$$Y = r_0 + r_1X_1 + r_2X_2 + \cdots + r_kX_k + \epsilon \tag{1.24}$$

then, regardless of the underlying distribution of $Y, X_1, X_2, \ldots, X_k$, the best least-square coefficients are obtained when ϵ is uncorrelated with each of the regressors $X_1, X_2, \ldots, X_k$. That is,

$$Cov(\epsilon, X_i) = 0 \quad \text{for} \quad i = 1,2, \ldots, k$$

To see how this *orthogonality principle* is used to our advantage, assume we wish to compute the best estimate of $X = Die\ 1$ given the sum

$$Y = Die\ 1 + Die\ 2$$

Writing

$$X = \alpha + \beta Y + \epsilon \tag{1.25a}$$

our goal is to find α and β in terms of estimable statistical measures. Assuming without loss of generality $E[\epsilon] = 0$, and taking expectation on both sides of the equation, we obtain

$$E[X] = \alpha + \beta E[Y] \tag{1.25b}$$

Further multiplying both sides of (1.25a) by Y and taking the expectation gives

$$E[XY] = \alpha E[Y] + \beta E[Y^2] + E[Y\epsilon] \tag{1.26}$$

The orthogonality principle dictates $E[Y\epsilon] = 0$, and (1.25b) and (1.26) yield two equations with two unknowns, α and β. Solving for α and β, we obtain

$$\alpha = E(X) - E(Y)\frac{\sigma_{XY}}{\sigma_Y^2}$$

$$\beta = \frac{\sigma_{XY}}{\sigma_Y^2}$$

which completes the derivation. The slope β could have been obtained from Eq. (1.22), by simply reversing X and Y, but the derivation above demonstrates a general method of computing slopes, in two or more dimensions.

Consider for example the problem of finding the best estimate of Z given two observations, $X = x$ and $Y = y$. As before, we write the regression equation

$$Z = \alpha + \beta_Y Y + \beta_X X + \epsilon$$

But now, to obtain three equations for α, β_Y, and β_X, we also multiply both sides by Y and X and take expectations. Imposing the orthogonality conditions $E[\epsilon Y] = E[\epsilon X] = 0$ and solving the resulting equations gives

$$\beta_Y = R_{ZY \cdot X} = \frac{\sigma_X^2 \sigma_{ZY} - \sigma_{ZX}\sigma_{XY}}{\sigma_Y^2 \sigma_X^2 - \sigma_{YX}^2} \tag{1.27}$$

$$\beta_X = R_{ZX \cdot Y} = \frac{\sigma_Y^2 \sigma_{ZX} - \sigma_{ZY}\sigma_{YX}}{\sigma_Y^2 \sigma_X^2 - \sigma_{YX}^2} \tag{1.28}$$

Equations (1.27) and (1.28) are generic; they give the linear regression coefficients $R_{ZY \cdot X}$ and $R_{ZX \cdot Y}$ for any three variables in terms of their variances and covariances, and as such, they allow us to see how sensitive these slopes are to other model parameters. In practice, however, regression slopes are estimated from sampled data by efficient "least-square" algorithms, and rarely require memorization of mathematical equations. An exception is the task of predicting whether any of these slopes is zero, prior to obtaining any data. Such predictions are important when we contemplate choosing a set of regressors for one purpose or another, and as we shall see in Section 3.8, this task will be handled quite efficiently through the use of causal graphs.

Study question 1.3.9

(a) Prove Eq. (1.22) using the orthogonality principle. [Hint: Follow the treatment of Eq. (1.26).]

(b) Find all partial regression coefficients

$$R_{YX \cdot Z}, R_{XY \cdot Z}, R_{YZ \cdot X}, R_{ZY \cdot X}, R_{XZ \cdot Y}, \text{ and } R_{ZX \cdot Y}$$

for the craps game described in Study question 1.3.8. [Hint: Apply Eq. (1.27) and use the variances and covariances computed for part (a) of Study question 1.3.8.]

1.4 Graphs

We learned from Simpson's Paradox that certain decisions cannot be made on the basis of data alone, but instead depend on the story behind the data. In this section, we layout a mathematical language, *graph theory*, in which these stories can be conveyed. Graph theory is not generally taught in high school mathematics, but it provides a useful mathematical language that allows us to address problems of causality with simple operations similar to those used to solve arithmetic problems.

Although the word *graph* is used colloquially to refer to a whole range of visual aids—more or less interchangeably with the word *chart*—in mathematics, a graph is a formally defined object. A mathematical graph is a collection of *vertices* (or, as we will call them, *nodes*) and edges. The nodes in a graph are connected (or not) by the edges. Figure 1.5 illustrates a simple graph. X, Y, and Z (the dots) are nodes, and A and B (the lines) are edges.

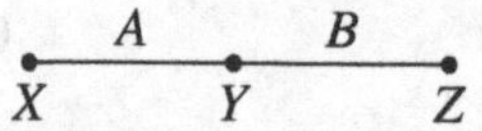

Figure 1.5 An undirected graph in which nodes X and Y are adjacent and nodes Y and Z are adjacent but not X and Z

Two nodes are *adjacent* if there is an edge between them. In Figure 1.5, X and Y are adjacent, and Y and Z are adjacent. A graph is said to be a *complete graph* if there is an edge between every pair of nodes in the graph.

A *path* between two nodes X and Y is a sequence of nodes beginning with X and ending with Y, in which each node is connected to the next by an edge. For instance, in Figure 1.5, there is a path from X to Z, because X is connected to Y, and Y is connected to Z.

Edges in a graph can be *directed* or *undirected.* Both of the edges in Figure 1.5 are undirected, because they have no designated "in" and "out" ends. A directed edge, on the other hand, goes out of one node and into another, with the direction indicated by an arrow head. A graph in which all of the edges are directed is a *directed graph.* Figure 1.6 illustrates a directed graph. In Figure 1.6, A is a directed edge from X to Y and B is a directed edge from Y to Z.

Figure 1.6 A directed graph in which node X is a parent of Y and Y is a parent of Z

The node that a directed edge starts from is called the *parent* of the node that the edge goes into; conversely, the node that the edge goes into is the *child* of the node it comes from. In Figure 1.6, X is the parent of Y, and Y is the parent of Z; accordingly, Y is the child of X, and Z is the child of Y. A path between two nodes is a *directed* path if it can be traced along the arrows, that is, if no node on the path has two edges on the path directed into it, or two edges directed out of it. If two nodes are connected by a directed path, then the first node is the *ancestor* of every node on the path (excluding itself), and the last node is the *descendant* of every node on the path (excluding itself). (Think of this as an analogy to parent nodes and child nodes: parents are the ancestors of their children, and of their children's children, and of their children's children's children, etc.) For instance, in Figure 1.6, X is the ancestor of both Y and Z, and both Y and Z are descendants of X.

When a directed path exists from a node to itself, the path (and graph) is called *cyclic.* A directed graph with no cycles is *acyclic.* For example, in Figure 1.7(a) the graph is acyclic; however, the graph in Figure 1.7(b) is cyclic. Note that in 1.7(a) there is no directed path from any node to itself, whereas in 1.7(b) there are directed paths from X back to X, for example.

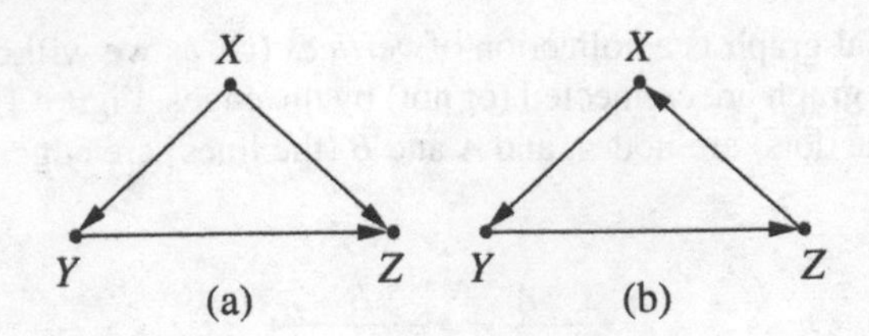

Figure 1.7 (a) Showing acyclic graph and (b) cyclic graph

Study questions

Study question 1.4.1

Consider the graph shown in Figure 1.8:

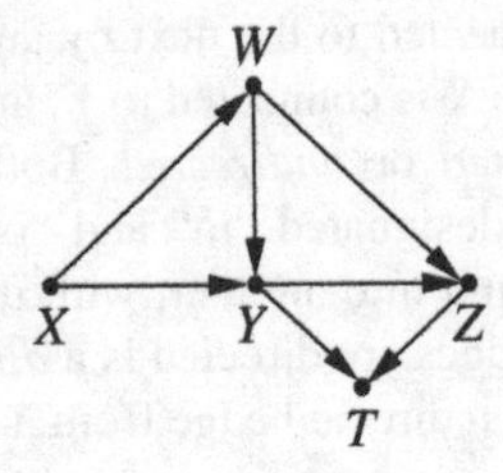

Figure 1.8 A directed graph used in Study question 1.4.1

(a) Name all of the parents of Z.
(b) Name all the ancestors of Z.
(c) Name all the children of W.
(d) Name all the descendants of W.
(e) Draw all (simple) paths between X and T (i.e., no node should appear more than once).
(f) Draw all the directed paths between X and T.

1.5 Structural Causal Models

1.5.1 Modeling Causal Assumptions

In order to deal rigorously with questions of causality, we must have a way of formally setting down our assumptions about the causal story behind a data set. To do so, we introduce the concept of the *structural causal model*, or SCM, which is a way of describing the relevant features of the world and how they interact with each other. Specifically, a structural causal model describes how nature assigns values to variables of interest.

Formally, a structural causal model consists of two sets of variables U and V, and a set of functions f that assigns each variable in V a value based on the values of the other variables in the model. Here, as promised, we expand on our definition of causation: A variable X is a *direct cause* of a variable Y if X appears in the function that assigns Y's value. X is a *cause* of Y if it is a direct cause of Y, or of any cause of Y.

The variables in U are called *exogenous variables*, meaning, roughly, that they are external to the model; we choose, for whatever reason, not to explain how they are caused. The variables in V are *endogenous*. Every endogenous variable in a model is a descendant of at least one exogenous variable. Exogenous variables cannot be descendants of any other variables, and in particular, cannot be a descendant of an endogenous variable; they have no ancestors and are represented as *root* nodes in graphs. If we know the value of every exogenous variable, then using the functions in f, we can determine with perfect certainty the value of every endogenous variable.

For example, suppose we are interested in studying the causal relationships between a treatment X and lung function Y for individuals who suffer from asthma. We might assume that Y also depends on, or is "caused by," air pollution levels as captured by a variable Z. In this case, we would refer to X and Y as endogenous and Z as exogenous. This is because we assume that air pollution is an external factor, that is, it cannot be caused by an individual's selected treatment or their lung function.

Every SCM is associated with a *graphical causal model*, referred to informally as a "graphical model" or simply "graph." Graphical models consist of a set of nodes representing the variables in U and V, and a set of edges between the nodes representing the functions in f. The graphical model G for an SCM M contains one node for each variable in M. If, in M, the function f_X for a variable X contains within it the variable Y (i.e., if X depends on Y for its value), then, in G, there will be a directed edge from Y to X. We will deal primarily with SCMs for which the graphical models are *directed acyclic graphs* (DAGs). Because of the relationship between SCMs and graphical models, we can give a graphical definition of causation: If, in a graphical model, a variable X is the child of another variable Y, then Y is a direct cause of X; if X is a descendant of Y, then Y is a potential cause of X (there are rare *intransitive cases* in which Y will not be a cause of X, which we will discuss in Part Two).

In this way, causal models and graphs encode causal assumptions. For instance, consider the following simple SCM:

SCM 1.5.1 (Salary Based on Education and Experience)

$$U = \{X, Y\}, \quad V = \{Z\}, \quad F = \{f_Z\}$$

$$f_Z : Z = 2X + 3Y$$

This model represents the salary (Z) that an employer pays an individual with X years of schooling and Y years in the profession. X and Y both appear in f_Z, so X and Y are both direct causes of Z. If X and Y had any ancestors, those ancestors would be potential causes of Z.

The graphical model associated with SCM 1.5.1 is illustrated in Figure 1.9.

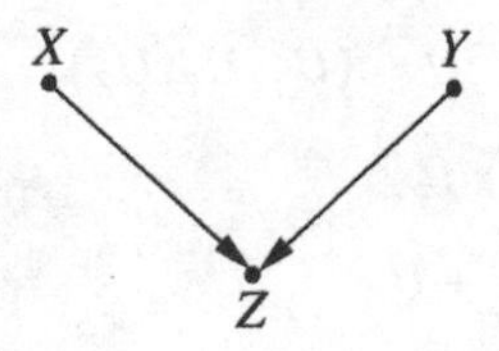

Figure 1.9 The graphical model of SCM 1.5.1, with X indicating years of schooling, Y indicating years of employment, and Z indicating salary

Because there are edges connecting Z to X and Y, we can conclude just by looking at the graphical model that there is some function f_Z in the model that assigns Z a value based on X and Y, and therefore that X and Y are causes of Z. However, without the fuller specification of an SCM, we can't tell from the graph what the function is that defines Z—or, in other words, *how* X and Y cause Z.

If graphical models contain less information than SCMs, why do we use them at all? There are several reasons. First, usually the knowledge that we have about causal relationships is not quantitative, as demanded by an SCM, but qualitative, as represented in a graphical model. We know off-hand that sex is a cause of height and that height is a cause of performance in basketball, but we would hesitate to give numerical values to these relationships. We could, instead of drawing a graph, simply create a partially specified version of the SCM:

SCM 1.5.2 (Basketball Performance Based on Height and Sex)

$$V = \{\text{Height, Sex, Performance}\}, \quad U = \{U_1, U_2, U_3\}, \quad F = \{f1, f2\}$$

$$\text{Sex} = U_1$$

$$\text{Height} = f_1(\text{Sex}, U_2)$$

$$\text{Performance} = f_2(\text{Height, Sex}, U_3)$$

Here, $U = \{U_1, U_2, U_3\}$ represents unmeasured factors that we do not care to name, but that affect the variables in V that we can measure. The U factors are sometimes called "error terms" or "omitted factors." These represent additional unknown and/or random exogenous causes of what we observe.

But graphical models provide a more intuitive understanding of causality than do such partially specified SCMs. Consider the SCM and its associated graphical model introduced above; while the SCM and its graphical model contain the same information, that is, that X causes Z and Y causes Z, that information is more quickly and easily ascertained by looking at the graphical model.

Study questions

Study question 1.5.1

Suppose we have the following SCM. Assume all exogenous variables are independent and that the expected value of each is 0.

SCM 1.5.3

$$V = \{X, Y, Z\}, \quad U = \{U_X, U_Y, U_Z\}, \quad F = \{f_X, f_Y, f_Z\}$$

$$f_X : X = U_X$$

$$f_Y : Y = \frac{X}{3} + U_Y$$

$$f_Z : Z = \frac{Y}{16} + U_Z$$

(a) *Draw the graph that complies with the model.*
(b) *Determine the best guess of the value (expected value) of Z, given that we observe $Y = 3$.*
(c) *Determine the best guess of the value of Z, given that we observe $X = 3$.*
(d) *Determine the best guess of the value of Z, given that we observe $X = 1$ and $Y = 3$.*
(e) *Assume that all exogenous variables are normally distributed with zero means and unit variance, that is, $\sigma = 1$.*
 (i) *Determine the best guess of X, given that we observed $Y = 2$.*
 (ii) *(Advanced) Determine the best guess of Y, given that we observed $X = 1$ and $Z = 3$. [Hint: You may wish to use the technique of multiple regression, together with the fact that, for every three normally distributed variables, say X, Y, and Z, we have $E[Y|X = x, Z = z] = R_{YX \cdot Z}x + R_{YZ \cdot X}z$.]*

1.5.2 Product Decomposition

Another advantage of graphical models is that they allow us to express joint distributions very efficiently. So far, we have presented joint distributions in two ways. First, we have used tables, in which we assigned a probability to every possible combination of values. This is intuitively easy to parse, but in models with many variables, it can take up a prohibitive amount of space; 10 binary variables would require a table with 1024 rows!

Second, in a fully specified SCM, we can represent the joint distributions of n variables with greater efficiency: We need only to specify the n functions that govern the relationships between the variables, and then from the probabilities of the error terms, we can discover all the probabilities that govern the joint distribution. But we are not always in a position to fully specify a model; we may know that one variable is a cause of another but not the form of the equation relating them, or we may not know the distributions of the error terms. Even if we know these objects, writing them down may be easier said than done, especially when the variables are discrete and the functions do not have familiar algebraic expressions.

Fortunately, we can use graphical models to help overcome both of these barriers through the following rule.

Rule of product decomposition

For any model whose graph is acyclic, the joint distribution of the variables in the model is given by the product of the conditional distributions *P(child|parents)* over all the "families" in the graph. Formally, we write this rule as

$$P(x_1, x_2, \dots, x_n) = \prod_i P(x_i|pa_i) \tag{1.29}$$

where pa_i stands for the values of the parents of variable X_i, and the product $\prod_i$ runs over all i, from 1 to n. The relationship (1.29) follows from certain universally true independencies among the variables, which will be discussed in the next chapter in more detail.

For example, in a simple chain graph $X \to Y \to Z$, we can write directly:

$$P(X = x, Y = y, Z = z) = P(X = x)P(Y = y|X = x)P(Z = z|Y = y)$$

This knowledge allows us to save an enormous amount of space when laying out a joint distribution. We need not create a probability table that lists a value for every possible triple (x, y, z). It will suffice to create three much smaller tables for X, $(Y|X)$, and $(Z|Y)$, and multiply the values as necessary.

To estimate the joint distribution from a data set generated by the above model, we need not count the frequency of every triple; we can instead count the frequencies of each x, $(y|x)$, and $(z|y)$ and multiply. This saves us a great deal of processing time in large models. It also increases substantially the accuracy of frequency counting. Thus, the assumptions underlying the graph allow us to exchange a "high-dimensional" estimation problem for a few "low-dimensional" probability distribution challenges. The graph therefore simplifies an estimation problem and, simultaneously, provides more precise estimators. If we do not know the graphical structure of an SCM, estimation becomes impossible with a large number of variables and small, or moderately sized, data sets—the so-called "curse of dimensionality."

Graphical models let us do all of this without always needing to know the functions relating the variables, their parameters, or the distributions of their error terms.

Here's an evocative, if unrigorous, demonstration of the time and space saved by this strategy: Consider the chain $X \rightarrow Y \rightarrow Z \rightarrow W$, where X stands for clouds/no clouds, Y stands for rain/no rain, Z stands for wet pavement/dry pavement, and W stands for slippery pavement/unslippery pavement.

Using your own judgment, based on your experience of the world, how plausible is it that *P(clouds, no-rain, dry pavement, slippery pavement)* = 0.23?

This is quite a difficult question to answer straight out. But using the product rule, we can break it into pieces:

P(clouds)P(no rain|clouds)P(dry pavement|no rain)P(slippery pavement|dry pavement)

Our general sense of the world tells us that *P(clouds)* should be relatively high, perhaps 0.5 (lower, of course, for those of us living in the strange, weatherless city of Los Angeles). Similarly, *P(no rain|clouds)* is fairly high—say, 0.75. And *P(dry pavement|no rain)* would be higher still, perhaps 0.9. But the *P(slippery pavement|dry pavement)* should be quite low, somewhere in the range of 0.05. So putting it all together, we come to a ballpark estimate of $0.5 \times 0.75 \times 0.9 \times 0.05 \approx 0.0169$.

We will use this product rule often in this book in cases when we need to reason with numerical probabilities, but wish to avoid writing out large probability tables.

The importance of the product decomposition rule can be particularly appreciated when we deal with estimation. In fact, much of the role of statistics focuses on effective sampling designs, and estimation strategies, that allow us to exploit an appropriate data set to estimate probabilities as precisely as we might need. Consider again the problem of estimating the probability $P(X, Y, Z, W)$ for the chain $X \rightarrow Y \rightarrow Z \rightarrow W$. This time, however, we attempt to estimate the probability from data, rather than our own judgment. The number of (x, y, z, w) combinations that need to be assigned probabilities is $16 - 1 = 15$ (the last probability can be computed from the sum of the others). Assume that we have 45 random observations, each consisting of a vector (x, y, z, w). On the average, each (x, y, z, w) cell would receive less than three samples; some will receive one or two samples, and some remain empty. It is very unlikely that we would obtain a sufficient number of samples in each cell to assess the proportion in the population at large (i.e., when the sample size goes to infinity).

If we use our product decomposition rule, however, the 45 samples are separated into much larger categories. In order to determine $P(x)$, every (x, y, z, w) sample falls into one of only two cells: $(X = 1)$ and $(X = 0)$. Clearly, the probability of leaving either of them empty is much lower, and the accuracy of estimating population frequencies is much higher. The same is true of the divisions we need to make to determine $P(y|x) : (Y = 1, X = 1), (Y = 0, X = 1), (Y = 1, X = 0)$, and $(Y = 0, X = 0)$. And to determine $P(z|y) : (Y = 1, Z = 1), (Y = 0, Z = 1), (Y = 1, Z = 0)$, and $(Y = 0, Z = 0)$. And to determine $P(w|z) : (W = 1, Z = 1), (W = 0, Z = 1), (W = 1, Z = 0)$, and $(W = 0, Z = 0)$. Each of these divisions will give us much more accurate frequencies than our original division into 16 cells. Here we explicitly see the simpler estimation problems allowed by assuming the graphical structure of an SCM and the resulting improved accuracy of our frequency estimates.

This is not the only use to which we can put the qualitative knowledge that a graph provides. As we will see in the next section, graphical models reveal much more information than is obvious at first glance; we can learn a lot about, and infer a lot from, a data set using only the graphical model of its causal story.

Study questions

Study question 1.5.2

Assume that a population of patients contains a fraction r of individuals who suffer from a certain fatal syndrome Z, which simultaneously makes it uncomfortable for them to take a life-prolonging drug X (Figure 1.10). Let $Z = z_1$ and $Z = z_0$ represent, respectively, the presence and absence of the syndrome, $Y = y_1$ and $Y = y_0$ represent death and survival, respectively, and $X = x_1$ and $X = x_0$ represent taking and not taking the drug. Assume that patients not carrying the syndrome, $Z = z_0$, die with probability p_2 if they take the drug and with probability p_1 if they don't. Patients carrying the syndrome, $Z = z_1$, on the other hand, die with probability p_3 if they do not take the drug and with probability p_4 if they do take the drug. Further, patients having the syndrome are more likely to avoid the drug, with probabilities $q_1 = P(x_1|z_0)$ and $q_2 = P(x_1|z_1)$.

(a) Based on this model, compute the joint distributions $P(x, y, z)$, $P(x, y)$, $P(x, z)$, and $P(y, z)$ for all values of x, y, and z, in terms of the parameters $(r, p_1, p_2, p_3, p_4, q_1, q_2)$. [Hint: Use the product decomposition of Section 1.5.2.]

(b) Calculate the difference $P(y_1|x_1) - P(y_1|x_o)$ for three populations: (1) those carrying the syndrome, (2) those not carrying the syndrome, and (3) the population as a whole.

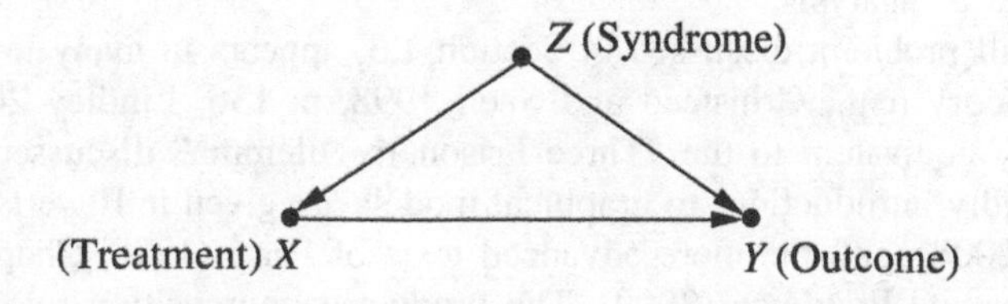

Figure 1.10 Model showing an unobserved syndrome, Z, affecting both treatment (X) and outcome (Y)

(c) Using your results for (b), find a combination of parameters that exhibits Simpson's reversal.

Study question 1.5.3

Consider a graph $X_1 \rightarrow X_2 \rightarrow X_3 \rightarrow X_4$ *of binary random variables, and assume that the conditional probabilities between any two consecutive variables are given by*

$$P(X_i = 1|X_{i-1} = 1) = p$$

$$P(X_i = 1|X_{i-1} = 0) = q$$

$$P(X_1 = 1) = p_0$$

Compute the following probabilities

$$P(X_1 = 1, X_2 = 0, X_3 = 1, X_4 = 0)$$

$$P(X_4 = 1|X_1 = 1)$$

$$P(X_1 = 1|X_4 = 1)$$

$$P(X_3 = 1|X_1 = 0, X_4 = 1)$$

Study question 1.5.4

Define the structural model that corresponds to the Monty Hall problem, and use it to describe the joint distribution of all variables.

Bibliographical Notes for Chapter 1

An extensive account of the history of Simpson's paradox is given in Pearl (2009, pp. 174–182), including many attempts by statisticians to resolve it without invoking causation. A more recent account, geared for statistics instructors is given in (Pearl 2014b). Among the many texts that provide basic introductions to probability theory, Lindley (2014) and Pearl (1988, Chapters 1 and 2) are the closest in spirit to the Bayesian perspective used in Chapter 1. The textbooks by Selvin (2004) and Moore et al. (2014) provide excellent introductions to classical methods of statistics, including parameter estimation, hypothesis testing and regression analysis.

The Monty Hall problem, discussed in Section 1.3, appears in many introductory books on probability theory (e.g., Grinstead and Snell 1998, p. 136; Lindley 2014, p. 201) and is mathematically equivalent to the "Three Prisoners Dilemma" discussed in (Pearl 1988, pp. 58–62). Friendly introductions to graphical models are given in Elwert (2013), Glymour and Greenland (2008), and the more advanced texts of Pearl (1988, Chapter 3), Lauritzen (1996) and Koller and Friedman (2009). The product decomposition rule of Section 1.5.2 was used in Howard and Matheson (1981) and Kiiveri et al. (1984) and became the semantic basis of *Bayesian Networks* (Pearl 1985)—directed acyclic graphs that represent probabilistic knowledge, not necessarily causal. For inference and applications of Bayesian networks, see Darwiche (2009) and Fenton and Neil (2013), and Conrady and Jouffe (2015). The validity of the product decomposition rule for structural causal models was shown in Pearl and Verma (1991).

2

Graphical Models and Their Applications

2.1 Connecting Models to Data

In Chapter 1, we treated probabilities, graphs, and structural equations as isolated mathematical objects with little to connect them. But the three are, in fact, closely linked. In this chapter, we show that the concept of independence, which in the language of probability is defined by algebraic equalities, can be expressed visually using directed acyclic graphs (DAGs). Further, this graphical representation will allow us to capture the probabilistic information that is implied by the structural equations.

The net result is that a researcher who has scientific knowledge in the form of a structural causal model is able to predict patterns of independencies in the data, based solely on the structure of the model's graph, without relying on any quantitative information carried by the equations or by the distributions of the errors. Conversely, it means that observing patterns of independencies in the data enables us to say something about whether a hypothesized model is correct. Ultimately, as we will see in Chapter 3, the structure of the graph, when combined with data, will enable us to predict quantitatively the results of interventions without actually performing them.

2.2 Chains and Forks

We have so far referred to causal models as representations of the "causal story" underlying data. Another way to think of this is that causal models represent the *mechanism* by which data were generated. Causal models are a sort of blueprint of the relevant part of the universe, and we can use them to simulate data from this universe. Given a truly complete causal model for, say, math test scores in high school juniors, and given a complete list of values for every exogenous variable in that model, we could theoretically generate a data point (i.e., a test score) for any individual. Of course, this would necessitate specifying all factors that may have an effect on a student's test score, an unrealistic task. In most cases, we will not have such precise knowledge about a model. We might instead have a probability distribution characterizing the exogenous

Causal Inference in Statistics: A Primer, First Edition. Judea Pearl, Madelyn Glymour, and Nicholas P. Jewell.

Companion Website: www.wiley.com/go/Pearl/Causality

variables, which would allow us to generate a distribution of test scores approximating that of the entire student population and relevant subgroups of students.

Suppose, however, that we do not have even a probabilistically specified causal model, but only a graphical structure of the model. We know which variables are caused by which other variables, but we don't know the strength or nature of the relationships. Even with such limited information, we can discern a great deal about the data set generated by the model. From an unspecified graphical causal model—that is, one in which we know which variables are functions of which others, but not the specific nature of the functions that connect them—we can learn which variables in the data set are independent of each other and which are independent of each other conditional on other variables. These independencies will be true of every data set generated by a causal model with that graphical structure, regardless of the specific functions attached to the SCM.

Consider, for instance, the following three hypothetical SCMs, all of which share the same graphical model. The first SCM represents the causal relationships among a high school's funding in dollars (*X*), its average SAT score (*Y*), and its college acceptance rate (*Z*) for a given year. The second SCM represents the causal relationships among the state of a light switch (*X*), the state of an associated electrical circuit (*Y*), and the state of a light bulb (*Z*). The third SCM concerns the participants in a foot race. It represents causal relationships among the hours that participants work at their jobs each week (*X*), the hours the participants put into training each week (*Y*), and the completion time, in minutes, the participants achieve in the race (*Z*). In all three models, the exogenous variables (U_X, U_Y, U_Z) stand in for any unknown or random effects that may alter the relationship between the endogenous variables. Specifically, in SCMs 2.2.1 and 2.2.3, U_Y and U_Z are additive factors that account for variations among individuals. In SCM 2.2.2, U_Y and U_Z take the value 1 if there is some unobserved abnormality, and 0 if there is none.

SCM 2.2.1 (School Funding, SAT Scores, and College Acceptance)

$$V = \{X, Y, Z\}, U = \{U_X, U_Y, U_Z\}, F = \{f_X, f_Y, f_Z\}$$

$$f_X : X = U_X$$

$$f_Y : Y = \frac{x}{3} + U_Y$$

$$f_Z : Z = \frac{y}{16} + U_Z$$

SCM 2.2.2 (Switch, Circuit, and Light Bulb)

$$V = \{X, Y, Z\}, U = \{U_X, U_Y, U_Z\}, F = \{f_X, f_Y, f_Z\}$$

$$f_X : X = U_X$$

$$f_Y : Y = \begin{cases} \text{Closed IF } (X = \text{Up AND } U_Y = 0) \text{ OR } (X = \text{Down AND } U_Y = 1) \\ \text{Open otherwise} \end{cases}$$

$$f_Z : Z = \begin{cases} \text{On IF } (Y = \text{Closed AND } U_Z = 0) \text{ OR } (Y = \text{Open AND } U_Z = 1) \\ \text{Off otherwise} \end{cases}$$

SCM 2.2.3 (Work Hours, Training, and Race Time)

$$V = \{X, Y, Z\}, U = \{U_X, U_Y, U_Z\}, F = \{f_X, f_Y, f_Z\}$$

$$f_X : X = U_X$$

$$f_Y : Y = 84 - x + U_Y$$

$$f_Z : Z = \frac{100}{y} + U_Z$$

SCMs 2.2.1–2.2.3 share the graphical model shown in Figure 2.1.

SCMs 2.2.1 and 2.2.3 deal with continuous variables; SCM 2.2.2 deals with categorical variables. The relationships between the variables in 2.2.1 are all positive (i.e., the higher the value of the parent variable, the higher the value of the child variable); the correlations between the variables in 2.2.3 are all negative (i.e., the higher the value of the parent variable, the lower the value of the child variable); the correlations between the variables in 2.2.2 are not linear at all, but logical. No two of the SCMs share any functions in common. But because they share a common graphical structure, the data sets generated by all three SCMs must share certain independencies—and we can predict those independencies simply by examining the graphical model in Figure 2.1. The independencies shared by data sets generated by these three SCMs, and the dependencies that are likely shared by all such SCMs, are these:

1. ***Z and Y are likely dependent***
   For some $z, y, P(Z = z|Y = y) \neq P(Z = z)$
2. ***Y and X are likely dependent***
   For some $y, x, P(Y = y|X = x) \neq P(Y = y)$
3. ***Z and X are likely dependent***
   For some $z, x, P(Z = z|X = x) \neq P(Z = z)$
4. ***Z and X are independent, conditional on Y***
   For all $x, y, z, P(Z = z|X = x, Y = y) = P(Z = z|Y = y)$

To understand why these independencies and dependencies hold, let's examine the graphical model. First, we will verify that any two variables with an edge between them are likely dependent. Remember that an arrow from one variable to another indicates that the first variable causes the second—that is, the value of the first variable is part of the function that determines the value of the second. Therefore, the second variable *depends* on the first for its value; there

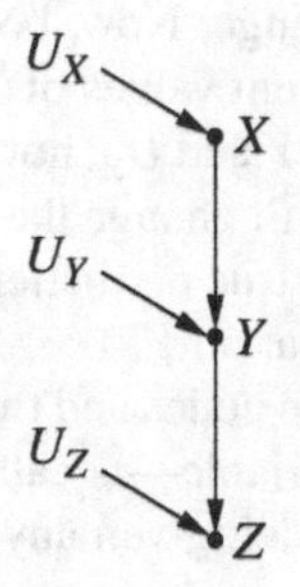

Figure 2.1 The graphical model of SCMs 2.2.1–2.2.3

is some case in which changing the value of the first variable changes the value of the second. That makes it likely that when we examine those variables in the data set, the probability that one variable takes a given value will change, given that we know the value of the other variable. So in a typical causal model, regardless of the specific functions, two variables connected by an edge are dependent. By this reasoning, we can see that in SCMs 2.2.1–2.2.3, Z and Y are likely dependent, and Y and X are likely dependent.[1]

From these two facts, we can conclude that Z and X are *likely* dependent. If Z depends on Y for its value, and Y depends on X for its value, then Z likely depends on X for its value. There are pathological cases in which this is not true. Consider, for example, the following SCM, which also has the graph in Figure 2.1.

SCM 2.2.4 (Pathological Case of Intransitive Dependence)

$$V = \{X, Y, Z\}, U = \{U_X, U_Y, U_Z\}, F = \{f_X, f_Y, f_Z\}$$

$$f_X : X = U_X$$

$$f_Y : Y = \begin{cases} a & \text{IF } X = 1 \text{ AND } U_Y = 1 \\ b & \text{IF } X = 2 \text{ AND } U_Y = 1 \\ c & \text{IF } U_Y = 2 \end{cases}$$

$$f_Z : Z = \begin{cases} i & \text{IF } Y = c \text{ OR } U_Z = 1 \\ j & \text{IF } Y \neq c \text{ AND } U_Z = 2 \end{cases}$$

In this case, no matter what value U_Y and U_Z take, X will have no effect on the value that Z takes; changes in X account for variation in Y between a and b, but Y doesn't affect Z unless it takes the value c. Therefore, X and Z vary independently in this model. We will call cases such as these *intransitive cases*.

However, intransitive cases form only a small number of the cases we will encounter. In most cases, the values of X and Z vary together just as X and Y do, and Y and Z. Therefore, they are likely dependent in the data set.

Now, let's consider point 4: Z and X are independent conditional on Y. Remember that when we condition on Y, we filter the data into groups based on the value of Y. So we compare all the cases where $Y = a$, all the cases where $Y = b$, and so on. Let's assume that we're looking at the cases where $Y = a$. We want to know whether, *in these cases only*, the value of Z is independent of the value of X. Previously, we determined that X and Z are likely dependent, because when the value of X changes, the value of Y likely changes, and when the value of Y changes, the value of Z is likely to change. Now, however, examining *only the cases where* $Y = a$, when we select cases with different values of X, the value of U_Y changes so as to keep Y at $Y = a$, but since Z depends only on Y and U_Z, not on U_Y, the value of Z remains unaltered. So selecting a different value of X doesn't change the value of Z. So, in the case where $Y = a$, X is independent of Z. This is of course true no matter which specific value of Y we condition on. So X is independent of Z, conditional on Y.

This configuration of variables—three nodes and two edges, with one edge directed into and one edge directed out of the middle variable—is called a *chain*. Analogous reasoning to the above tells us that in any graphical model, given any two variables X and Y, if the only path

[1] The independency can occur, for example, when X and U_Y are fair coins and $Y = 1$ if and only if $X = U_Y$. In this case, $P(Y = 1|X = 1) = P(Y = 1|X = 0) = P(Y = 1) = 1/2$. Such pathological cases require precise numerical probabilities to achieve independence ($P(X = 1) = P(U_Y = 1) = 1/2$); they are rare, and can be ignored for all practical purposes.

between X and Y is composed entirely of chains, then X and Y are independent conditional on any intermediate variable on that path. This independence relation holds regardless of the functions that connect the variables. This gives us a rule:

Rule 1 (Conditional Independence in Chains) *Two variables, X and Y, are conditionally independent given Z, if there is only one unidirectional path between X and Y and Z is any set of variables that intercepts that path.*

An important note: Rule 1 only holds when we assume that the error terms U_X, U_Y, and U_Z are independent of each other. If, for instance, U_X were a cause of U_Y, then conditioning on Z would not necessarily make X and Y independent—because variations in X could still be associated with variations in Y, through their error terms.

Now, consider the graphical model in Figure 2.2. This structure might represent, for example, the causal mechanism that connects a day's temperature in a city in degrees Fahrenheit (X), the number of sales at a local ice cream shop on that day (Y), and the number of violent crimes in the city on that day (Z). Possible functional relationships between these variables are given in SCM 2.2.5. Or the structure might represent, as in SCM 2.2.6, the causal mechanism that connects the state (up or down) of a switch (X), the state (on or off) of one light bulb (Y), and the state (on or off) of a second light bulb (Z). The exogenous variables U_X, U_Y, and U_Z represent other, possibly random, factors that influence the operation of these devices.

SCM 2.2.5 (Temperature, Ice Cream Sales, and Crime)

$$V = \{X, Y, Z\}, U = \{U_X, U_Y, U_Z\}, F = \{f_X, f_Y, f_Z\}$$

$$f_X : X = U_X$$

$$f_Y : Y = 4x + U_Y$$

$$f_Z : Z = \frac{x}{10} + U_Z$$

SCM 2.2.6 (Switch and Two Light Bulbs)

$$V = \{X, Y, Z\}, U = \{U_X, U_Y, U_Z\}, F = \{f_X, f_Y, f_Z\}$$

$$f_X : X = U_X$$

$$f_Y : Y = \begin{cases} \text{On IF } (X = \text{Up AND } U_Y = 0) \text{ OR } (X = \text{Down AND } U_Y = 1) \\ \text{Off otherwise} \end{cases}$$

$$f_Z : Z = \begin{cases} \text{On IF } (X = \text{Up AND } U_Z = 0) \text{ OR } (X = \text{Down AND } U_Z = 1) \\ \text{Off otherwise} \end{cases}$$

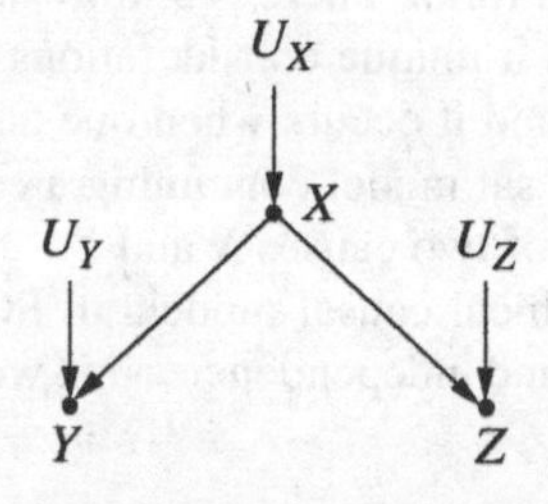

Figure 2.2 The graphical model of SCMs 2.2.5 and 2.2.6

If we assume that the error terms U_X, U_Y, and U_Z are independent, then by examining the graphical model in Figure 2.2, we can determine that SCMs 2.2.5 and 2.2.6 share the following dependencies and independencies:

1. ***X and Y are likely dependent.***
   For some $x, y, P(X = x|Y = y) \neq P(X = x)$
2. ***X and Z are likely dependent.***
   For some $x, z, P(X = x|Z = z) \neq P(X = x)$
3. ***Z and Y are likely dependent.***
   For some $z, y, P(Z = z|Y = y) \neq P(Z = z)$
4. ***Y and Z are independent, conditional on X.***
   For all $x, y, z, P(Y = y|Z = z, X = x) = P(Y = y|X = x)$

Points 1 and 2 follow, once again, from the fact that Y and Z are both directly connected to X by an arrow, so when the value of X changes, the values of both Y and Z likely change. This tells us something further, however: If Y changes when X changes, and Z changes when X changes, then it is likely (though not certain) that Y changes together with Z, and vice versa. Therefore, since a change in the value of Y gives us information about an associated change in the value of Z, Y and Z are likely dependent variables.

Why, then, are Y and Z independent conditional on X? Well, what happens when we condition on X? We filter the data based on the value of X. So now, we're only comparing cases where the value of X is constant. Since X does not change, the values of Y and Z do not change in accordance with it—they change only in response to U_Y and U_Z, which we have assumed to be independent. Therefore, any additional changes in the values of Y and Z must be independent of each other.

This configuration of variables—three nodes, with two arrows emanating from the middle variable—is called a *fork*. The middle variable in a fork is the *common cause* of the other two variables, and of any of their descendants. If two variables share a common cause, and if that common cause is part of the only path between them, then analogous reasoning to the above tells us that these dependencies and conditional independencies are true of those variables. Therefore, we come by another rule:

Rule 2 (Conditional Independence in Forks) *If a variable X is a common cause of variables Y and Z, and there is only one path between Y and Z, then Y and Z are independent conditional on X.*

2.3 Colliders

So far we have looked at two simple configurations of edges and nodes that can occur on a path between two variables: chains and forks. There is a third such configuration that we speak of separately, because it carries with it unique considerations and challenges. The third configuration contains a *collider* node, and it occurs when one node receives edges from two other nodes. The simplest graphical causal model containing a collider is illustrated in Figure 2.3, representing a common effect, Z, of two causes X and Y.

As is the case with every graphical causal model, all SCMs that have Figure 2.3 as their graph share a set of dependencies and independencies that we can determine from the graphical

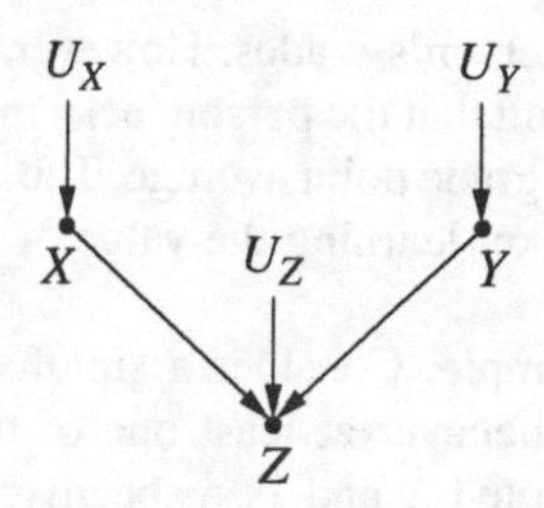

Figure 2.3 A simple collider

model alone. In the case of the model in Figure 2.3, assuming independence of U_X, U_Y, and U_Z, these independencies are as follows:

1. ***X and Z are likely dependent.***
   For some x, z, $P(X = x|Z = z) \neq P(X = x)$
2. ***Y and Z are likely dependent.***
   For some y, z, $P(Y = y|Z = z) \neq P(Y = y)$
3. ***X and Y are independent.***
   For all x, y, $P(X = x|Y = y) = P(X = x)$
4. ***X and Y are likely dependent conditional on Z.***
   For some x, y, z, $P(X = x|Y = y, Z = z) \neq P(X = x|Z = z)$

The truth of the first two points was established in Section 2.2. Point 3 is self-evident; neither X nor Y is a descendant or an ancestor of the other, nor do they depend for their value on the same variable. They respond only to U_X and U_Y, which are assumed independent, so there is no causal mechanism by which variations in the value of X should be associated with variations in the value of Y. This independence also reflects our understanding of how causation operates in time; events that are independent in the present do not become dependent merely because they may have common effects in the future.

Why, then, does point 4 hold? Why would two independent variables suddenly become dependent when we condition on their common effect? To answer this question, we return again to the definition of conditioning as filtering by the value of the conditioning variable. When we condition on Z, we limit our comparisons to cases in which Z takes the same value. But remember that Z depends, for its value, on X and Y. So, when comparing cases where Z takes some value, any change in value of X must be compensated for by a change in the value of Y—otherwise, the value of Z would change as well.

The reasoning behind this attribute of colliders—that conditioning on a collision node produces a dependence between the node's parents—can be difficult to grasp at first. In the most basic situation where $Z = X + Y$, and X and Y are independent variables, we have the following logic: If I tell you that $X = 3$, you learn nothing about the potential value of Y, because the two numbers are independent. On the other hand, if I start by telling you that $Z = 10$, then telling you that $X = 3$ immediately tells you that Y must be 7. Thus, X and Y are dependent, *given that* $Z = 10$.

This phenomenon can be further clarified through a real-life example. For instance, suppose a certain college gives scholarships to two types of students: those with unusual musical talents and those with extraordinary grade point averages. Ordinarily, musical talent and scholastic achievement are independent traits, so, in the population at large, finding a person with musical talent tells us nothing about that person's grades. However, discovering that a person is on a scholarship changes things; knowing that the person lacks musical talent then tells us immediately that he is likely to have a high grade point average. Thus, two variables that are marginally independent become dependent upon learning the value of a third variable (scholarship) that is a common effect of the first two.

Let's examine a numerical example. Consider a simultaneous (independent) toss of two fair coins and a bell that rings whenever at least one of the coins lands on heads. Let the outcomes of the two coins be denoted X and Y, respectively, and let Z stand for the state of the bell, with $Z = 1$ representing ringing, and $Z = 0$ representing silence. This mechanism can be represented as a collider as in Figure 2.3, in which the outcomes of the two coins are the parent nodes, and the state of the bell is the collision node.

If we know that Coin 1 landed on heads, it tells us nothing about the outcome of Coin 2, due to their independence. But suppose that we hear the bell ring and then we learn that Coin 1 landed on tails. We now know that Coin 2 must have landed on heads. Similarly, if we assume that we've heard the bell ring, the probability that Coin 1 landed on heads changes if we learn that Coin 2 also landed on heads. This particular change in probability is somewhat subtler than the first case.

To see the latter calculation, consider the initial probabilities as shown in Table 2.1.

We see that

$$P(X = \text{“Heads”}|Y = \text{“Heads”}) = P(X = \text{“Heads”}|Y = \text{“Tails”}) = \frac{1}{2}$$

That is, X and Y are independent. Now, let's condition on $Z = 1$ and $Z = 0$ (the bell ringing and not ringing). The resulting data subsets are shown in Table 2.2.

By calculating the probabilities in these tables, we obtain

$$P(X = \text{“Heads”}|Z = 1) = \frac{1}{3} + \frac{1}{3} = \frac{2}{3}$$

If we further filter the $Z = 1$ subtable to examine only those cases where $Y =$ *“Heads”*, we get

$$P(X = \text{“Heads”}|Y = \text{“Heads”}, Z = 1) = \frac{1}{2}$$

We see that, given $Z = 1$, the probability of $X =$ *“Heads”* changes from $\frac{2}{3}$ to $\frac{1}{2}$ upon learning that $Y =$ *“Heads.”* So, clearly, X and Y are dependent given $Z = 1$. A more pronounced dependence occurs, of course, when the bell does not ring ($Z = 0$), because then we know that both coins must have landed on tails.

Table 2.1 Probability distribution for two flips of a fair coin, with X representing flip one, Y representing flip two, and Z representing a bell that rings if either flip results in heads

X	Y	Z	$P(X, Y, Z)$
Heads	Heads	1	0.25
Heads	Tails	1	0.25
Tails	Heads	1	0.25
Tails	Tails	0	0.25

Table 2.2 Conditional probability distributions for the distribution in Table 2.1. (Top: Distribution conditional on $Z = 1$. Bottom: Distribution conditional on $Z = 0$)

X	Y	$P(X, Y \mid Z = 1)$
Heads	Heads	0.333
Heads	Tails	0.333
Tails	Heads	0.333
Tails	Tails	0
X	Y	$Pr(X, Y \mid Z = 0)$
Heads	Heads	0
Heads	Tails	0
Tails	Heads	0
Tails	Tails	1

Another example of colliders in action—one that may serve to further illuminate the difficulty that such configurations can present to statisticians—is the Monty Hall Problem, which we first encountered in Section 1.3. At its heart, the Monty Hall Problem reflects the presence of a collider. Your initial choice of door is one parent node; the door behind which the car is placed is the other parent node; and the door Monty opens to reveal a goat is the collision node, causally affected by both the other two variables. The causation here is clear: If you choose Door A, and if Door A has a goat behind it, Monty is forced to open whichever of the remaining doors that has a goat behind it.

Your initial choice and the location of the car are independent; that's why you initially have a $\frac{1}{3}$ chance of choosing the door with the car behind it. However, as with the two independent coins, conditional on Monty's choice of door, your initial choice and the placement of the prizes are dependent. Though the car may only be behind Door B in $\frac{1}{3}$ of cases, it will be behind Door B in $\frac{2}{3}$ of cases in which you choose Door A and Monty opened Door C.

Just as conditioning on a collider makes previously independent variables dependent, so too does conditioning on any descendant of a collider. To see why this is true, let's return to our example of two independent coins and a bell. Suppose we do not hear the bell directly, but instead rely on a witness who is somewhat unreliable; whenever the bell *does not ring*, there is 50% chance that our witness will falsely report that it did. Letting W stand for the witness's report, the causal structure is shown in Figure 2.4, and the probabilities for all combinations of X, Y, and W are shown in Table 2.3.

The reader can easily verify that, based on this table, we have

$$P(X = \text{“Heads”} \mid Y = \text{“Heads”}) = P(X = \text{“Heads”}) = \frac{1}{2}$$

and

$$P(X = \text{“Heads”} \mid W = 1) = (0.25 + 0.25) \div (0.25 + 0.25 + 0.25 + 0.125) = \frac{0.5}{0.875}$$

and

$$P(X = \text{“Heads”} \mid Y = \text{“Heads”}, W = 1) = 0.25 \div (0.25 + 0.25) = 0.5 < \frac{0.5}{0.875}$$

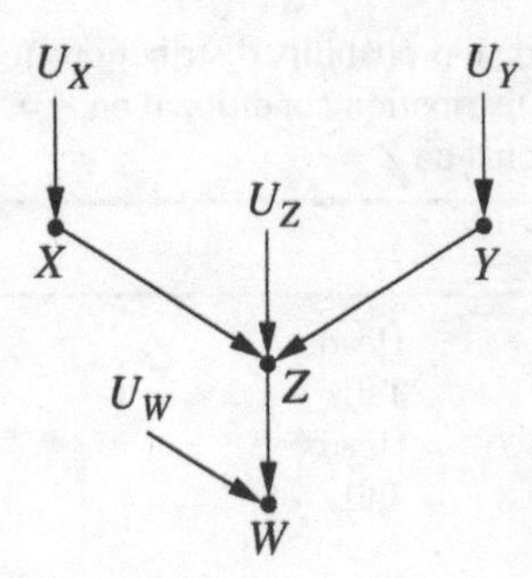

Figure 2.4 A simple collider, Z, with one child, W, representing the scenario from Table 2.3, with X representing one coin flip, Y representing the second coin flip, Z representing a bell that rings if either X or Y is heads, and W representing an unreliable witness who reports on whether or not the bell has rung

Table 2.3 Probability distribution for two flips of a fair coin and a bell that rings if either flip results in heads, with X representing flip one, Y representing flip two, and W representing a witness who, with variable reliability, reports whether or not the bell has rung

X	Y	W	$P(X, Y, W)$
Heads	Heads	1	0.25
Heads	Tails	1	0.25
Tails	Heads	1	0.25
Tails	Tails	1	0.125
Tails	Tails	0	0.125

Thus, X and Y are independent before reading the witness report, but become dependent thereafter.

These considerations lead us to a third rule, in addition to the two we established in Section 2.2.

Rule 3 (Conditional Independence in Colliders) *If a variable Z is the collision node between two variables X and Y, and there is only one path between X and Y, then X and Y are unconditionally independent but are dependent conditional on Z and any descendants of Z.*

Rule 3 is extremely important to the study of causality. In the coming chapters, we will see that it allows us to test whether a causal model could have generated a data set, to discover models from data, and to fully resolve Simpson's Paradox by determining which variables to measure and how to estimate causal effects under confounding.

Remark Inquisitive students may wonder why it is that dependencies associated with conditioning on a collider are so surprising to most people—as in, for example, the Monty Hall example. The reason is that humans tend to associate dependence with causation. Accordingly, they assume (wrongly) that statistical dependence between two variables can only exist if there is a causal mechanism that generates such dependence; that is, either one of the variables causes the other or a third variable causes both. In the case of a collider, they are surprised to find a

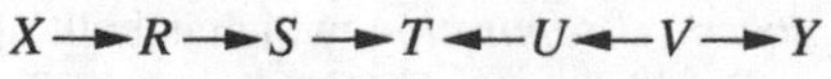

$$X \rightarrow R \rightarrow S \rightarrow T \leftarrow U \leftarrow V \rightarrow Y$$

Figure 2.5 A directed graph for demonstrating conditional independence (error terms are not shown explicitly)

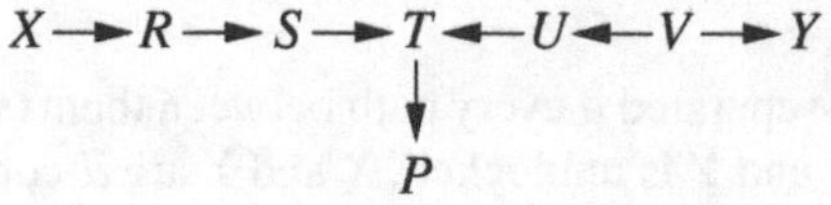

$$X \rightarrow R \rightarrow S \rightarrow T \leftarrow U \leftarrow V \rightarrow Y, \quad T \rightarrow P$$

Figure 2.6 A directed graph in which P is a descendant of a collider

dependence that is created in a third way, thus violating the assumption of "no correlation without causation."

Study questions

Study question 2.3.1

(a) List all pairs of variables in Figure 2.5 that are independent conditional on the set $Z = \{R, V\}$.

(b) For each pair of nonadjacent variables in Figure 2.5, give a set of variables that, when conditioned on, renders that pair independent.

(c) List all pairs of variables in Figure 2.6 that are independent conditional on the set $Z = \{R, P\}$.

(d) For each pair of nonadjacent variables in Figure 2.6, give a set of variables that, when conditioned on, renders that pair independent.

(e) Suppose we generate data by the model described in Figure 2.5, and we fit them with the linear equation $Y = a + bX + cZ$. Which of the variables in the model may be chosen for Z so as to guarantee that the slope b would be equal to zero? [Hint: Recall, a non zero slope implies that Y and X are dependent given Z.]

(f) Continuing question (e), but now in reference to Figure 2.6, suppose we fit the data with the equation:

$$Y = a + bX + cR + dS + eT + fP$$

which of the coefficients would be zero?

2.4 *d*-separation

Causal models are generally not as simple as the cases we have examined so far. Specifically, it is rare for a graphical model to consist of a single path between variables. In most graphical models, pairs of variables will have multiple possible paths connecting them, and each path will traverse a variety of chains, forks, and colliders. The question remains whether there is a criterion or process that can be applied to a graphical causal model of *any* complexity in order to predict dependencies that are shared by all data sets generated by that graph.

There is, indeed, such a process: *d-separation*, which is built upon the rules established in the previous section. d-separation (the d stands for "directional") allows us to determine, for any pair of nodes, whether the nodes are *d-connected*, meaning there exists a connecting path between them, or *d-separated*, meaning there exists no such path. When we say that a pair of nodes are d-separated, we mean that the variables they represent are definitely independent; when we say that a pair of nodes are *d-connected*, we mean that they are possibly, or most likely, dependent.[2]

Two nodes X and Y are d-separated if every path between them (should any exist) is *blocked*. If even one path between X and Y is unblocked, X and Y are d-connected. The paths between variables can be thought of as pipes, and dependence as the water that flows through them; if even one pipe is unblocked, some water can pass from one place to another, and if a single path is clear, the variables at either end will be dependent. However, a pipe need only be blocked in one place to stop the flow of water through it, and similarly, it takes only one node to block the passage of dependence in an entire path.

There are certain kinds of nodes that can block a path, depending on whether we are performing unconditional or conditional d-separation. If we are not conditioning on any variable, then only colliders can block a path. The reasoning for this is fairly straightforward: as we saw in Section 2.3, unconditional dependence can't pass through a collider. So if every path between two nodes X and Y has a collider in it, then X and Y cannot be unconditionally dependent; they must be marginally independent.

If, however, we are conditioning on a set of nodes Z, then the following kinds of nodes can block a path:

- A collider that is not conditioned on (i.e., not in Z), and that has no descendants in Z.
- A chain or fork whose middle node is in Z.

The reasoning behind these points goes back to what we learned in Sections 2.2 and 2.3. A collider does not allow dependence to flow between its parents, thus blocking the path. But Rule 3 tells us that when we condition on a collider or its descendants, the parent nodes may become dependent. So a collider whose collision node is not in the conditioning set Z would block dependence from passing through a path, but one whose collision node, or its descendants, *is* in the conditioning set would not. Conversely, dependence can pass through noncolliders—chains and forks—but Rules 1 and 2 tell us that when we condition on them, the variables on either end of those paths become independent (when we consider one path at a time). So any noncollision node in the conditioning set would block dependence, whereas one that is not in the conditioning set would allow dependence through.

We are now prepared to give a general definition of d-separation:

Definition 2.4.1 (d-separation) *A path p is blocked by a set of nodes Z if and only if*

1. *p contains a chain of nodes $A \rightarrow B \rightarrow C$ or a fork $A \leftarrow B \rightarrow C$ such that the middle node B is in Z (i.e., B is conditioned on), or*
2. *p contains a collider $A \rightarrow B \leftarrow C$ such that the collision node B is not in Z, and no descendant of B is in Z.*

[2] The d-connected variables will be dependent for almost all sets of functions assigned to arrows in the graph, the exception being the sorts of intransitive cases discussed in Section 2.2.

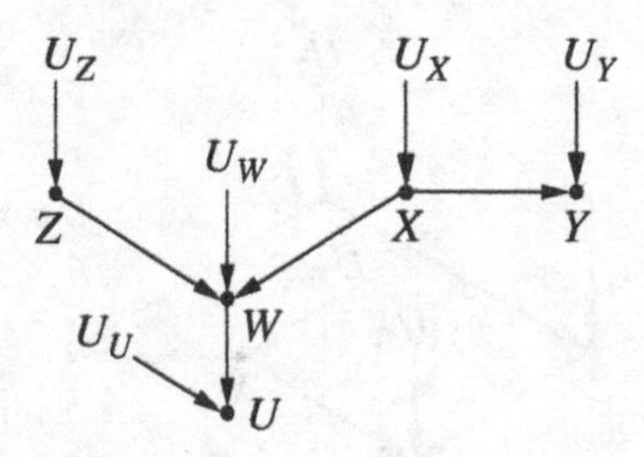

Figure 2.7 A graphical model containing a collider with child and a fork

If Z blocks every path between two nodes X and Y, then X and Y are d-separated, conditional on Z, and thus are independent conditional on Z.

Armed with the tool of d-separation, we can now look at some more complex graphical models and determine which variables in them are independent and dependent, both marginally and conditional on other variables. Let's take, for example, the graphical model in Figure 2.7. This graph might be associated with any number of causal models. The variables might be discrete, continuous, or a mixture of the two; the relationships between them might be linear, exponential, or any of an infinite number of other relations. No matter the model, however, d-separation will always provide the same set of independencies in the data the model generates.

In particular, let's look at the relationship between Z and Y. Using an empty conditioning set, they are d-separated, which tells us that Z and Y are unconditionally independent. Why? Because there is no unblocked path between them. There is only one path between Z and Y, and that path is blocked by a collider ($Z \rightarrow W \leftarrow X$).

But suppose we condition on W. d-separation tells us that Z and Y are d-connected, conditional on W. The reason is that our conditioning set is now $\{W\}$, and since the only path between Z and Y contains a fork (X) that is not in that set, and the only collider (W) on the path is in that set, that path is not blocked. (Remember that conditioning on colliders "unblocks" them.) The same is true if we condition on U, because U is a descendant of a collider along the path between Z and Y.

On the other hand, if we condition on the set $\{W, X\}$, Z and Y remain independent. This time, the path between Z and Y is blocked by the first criterion, rather than the second: There is now a noncollider node (X) on the path that is in the conditioning set. Though W has been unblocked by conditioning, one blocked node is sufficient to block the entire path. Since the only path between Z and Y is blocked by this conditioning set, Z and Y are d-separated conditional on $\{W, X\}$.

Now, consider what happens when we add another path between Z and Y, as in Figure 2.8. Z and Y are now unconditionally dependent. Why? Because there is a path between them ($Z \leftarrow T \rightarrow Y$) that contains no colliders. If we condition on T, however, that path is blocked, and Z and Y become independent again. Conditioning on $\{T, W\}$, on the other hand, makes them d-connected again (conditioning on T blocks the path $Z \leftarrow T \rightarrow Y$, but conditioning on W unblocks the path $Z \rightarrow W \leftarrow X \rightarrow Y$). And if we add X to the conditioning set, making it $\{T, W, X\}$, Z and Y become independent yet again! In this graph, Z and Y are d-connected (and therefore likely dependent) conditional

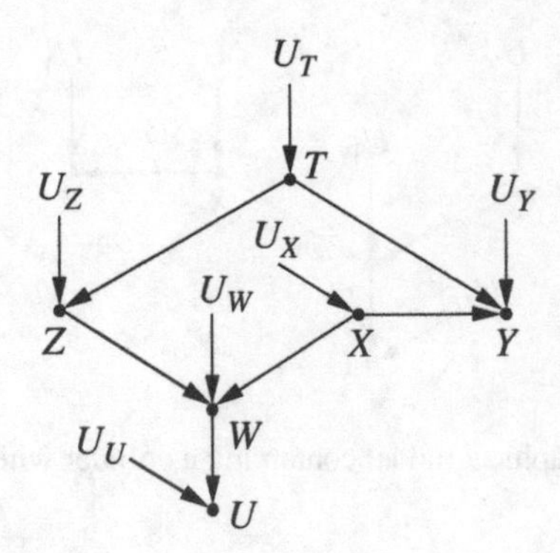

Figure 2.8 The model from Figure 2.7 with an additional forked path between Z and Y

on $W, U, X, \{W, U\}, \{W, T\}, \{U, T\}, \{W, U, T\}, \{W, X\}, \{U, X\}$, and $\{W, U, X\}$. They are d-separated (and therefore independent) conditional on $T, \{X, T\}, \{W, X, T\}, \{U, X, T\}$, and $\{W, U, X, T\}$. Note that T is in every conditioning set that d-separates Z and Y; that's because T is the only node in a path that unconditionally d-connects Z and Y, so unless it is conditioned on, Z and Y will always be d-connected.

Study questions

Study question 2.4.1

Figure 2.9 below represents a causal graph from which the error terms have been deleted. Assume that all those errors are mutually independent.

(a) *For each pair of nonadjacent nodes in this graph, find a set of variables that d-separates that pair. What does this list tell us about independencies in the data?*
(b) *Repeat question (a) assuming that only variables in the set $\{Z_3, W, X, Z_1\}$ can be measured.*
(c) *For each pair of nonadjacent nodes in the graph, determine whether they are independent conditional on all other variables.*
(d) *For every variable V in the graph, find a minimal set of nodes that renders V independent of all other variables in the graph.*
(e) *Suppose we wish to estimate the value of Y from measurements taken on all other variables in the model. Find the smallest set of variables that would yield as good an estimate of Y as when we measured all variables.*
(f) *Repeat question (e) assuming that we wish to estimate the value of Z_2.*
(g) *Suppose we wish to predict the value of Z_2 from measurements of Z_3. Would the quality of our prediction improve if we add measurement of W? Explain.*

2.5 Model Testing and Causal Search

The preceding sections demonstrate that causal models have *testable implications* in the data sets they generate. For instance, if we have a graph G that we believe might have generated

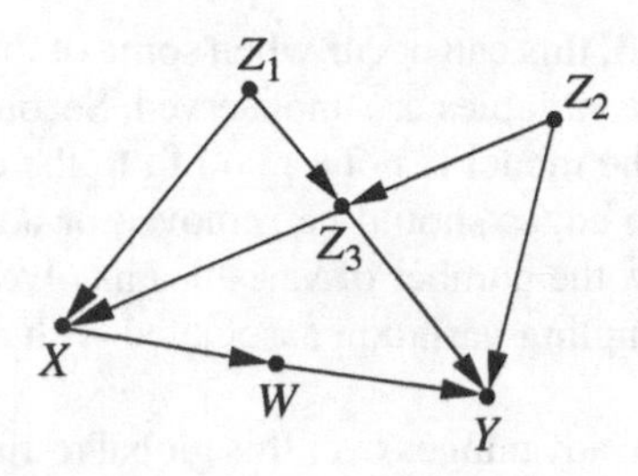

Figure 2.9 A causal graph used in study question 2.4.1, all U terms (not shown) are assumed independent

a data set S, d-separation will tell us which variables in G must be independent conditional on which other variables. Conditional independence is something we can test for using a data set. Suppose we list the d-separation conditions in G, and note that variables A and B must be independent conditional on C. Then, suppose we estimate the probabilities based on S, and discover that the data suggests that A and B are *not* independent conditional on C. We can then reject G as a possible causal model for S.

We can demonstrate it on the causal model of Figure 2.9. Among the many conditional independencies advertised by the model, we find that W and Z_1 are independent given X, because X d-separates W from Z_1. Now suppose we regress W on X and Z_1. Namely, we find the line

$$w = r_X x + r_1 z_1$$

that best fits our data. If it turns out that r_1 is not equal to zero, we know that W depends on Z_1 given X and, consequently, that the model is wrong. [Recall, conditional correlation implies conditional dependence.] Not only do we know that the model is wrong, but we also know where it is wrong; the true model must have a path between W and Z_1 that is not d-separated by X. Finally, this is a theoretical result that holds for all acyclic models with independent errors (Verma and Pearl 1990), and we also know that if every d-separation condition in the model matches a conditional independence in the data, then no further test can refute the model. This means that, for any data set whatsoever, one can always find a set of functions F for the model and an assignment of probabilities to the U terms, so as to generate the data precisely.

There are other methods for testing the fitness of a model. The standard way of evaluating fitness involves a statistical hypothesis test over the entire model, that is, we evaluate how likely it is for the observed samples to have been generated by the hypothesized model, as opposed to sheer chance. However, since the model is not fully specified, we need to first estimate its parameters before evaluating that likelihood. This can be done (approximately) when we assume a linear and Gaussian model (i.e., all functions in the model are linear and all error terms are normally distributed), because, under such assumptions, the joint distribution (also Gaussian) can be expressed succinctly in terms of the model's parameters, and we can then evaluate the likelihood that the observed samples have been generated by the fully parameterized model (Bollen 1989).

There are, however, a number of issues with this procedure. First, if any parameter cannot be estimated, then the joint distribution cannot be estimated, and the model cannot be tested.

As we shall see in Section 3.8.3, this can occur when some of the error terms are correlated or, equivalently, when some of the variables are unobserved. Second, this procedure tests models globally. If we discover that the model is not a good fit to the data, there is no way for us to determine why that is—which edges should be removed or added to improve the fit. Third, when we test a model globally, the number of variables involved may be large, and if there is measurement noise and/or sampling variation associated with each variable, the test will not be reliable.

d-separation presents several advantages over this global testing method. First, it is nonparametric, meaning that it doesn't rely on the specific functions that connect variables; instead, it uses only the graph of the model in question. Second, it tests models locally, rather than globally. This allows us to identify specific areas, where our hypothesized model is flawed, and to repair them, rather than starting from scratch on a whole new model. It also means that if, for whatever reason, we can't identify the coefficient in one area of the model, we can still get some incomplete information about the rest of the model. (As opposed to the first method, in which if we could not estimate one coefficient, we could not test any part of the model.)

If we had a computer, we could test and reject many possible models in this way, eventually whittling down the set of possible models to only a few whose testable implications do not contradict the dependencies present in the data set. It is a set of models, rather than a single model, because some graphs have indistinguishable implications. A set of graphs with indistinguishable implications is called an *equivalence class*. Two graphs G_1 and G_2 are in the same equivalence class if they share a common skeleton—that is, the same edges, regardless of the direction of those edges—and if they share common *v-structures*, that is, colliders whose parents are not adjacent. Any two graphs that satisfy this criterion have identical sets of *d*-separation conditions and, therefore, identical sets of testable implications (Verma and Pearl 1990).

The importance of this result is that it allows us to search a data set for the causal models that could have generated it. Thus, not only can we start with a causal model and generate a data set—but we can also start with a data set, and reason back to a causal model. This is enormously useful, since the object of most data-driven research is exactly to find a model that explains the data.

There are other methods of causal search—including some that rely on the kind of global model testing with which we began the section—but a full investigation of them is beyond the scope of this book. Those interested in learning more about search should refer to (Pearl 2000; Pearl and Verma 1991; Rebane and Pearl 1987; Spirtes and Glymour 1991; Spirtes et al. 1993).

Study questions

Study question 2.5.1

(a) Which of the arrows in Figure 2.9 can be reversed without being detected by any statistical test? [Hint: Use the criterion for equivalence class.]

(b) List all graphs that are observationally equivalent to the one in Figure 2.9.

(c) List the arrows in Figure 2.9 whose directionality can be determined from nonexperimental data.

(d) Write down a regression equation for Y such that, if a certain coefficient in that equation is nonzero, the model of Figure 2.9 is wrong.
(e) Repeat question (d) for variable Z_3.
(f) Repeat question (e) assuming the X is not measured.
(g) How many regression equations of the type described in (d) and (e) are needed to ensure that the model is fully tested, namely, that if it passes all these tests, it cannot be refuted by additional tests of these kind. [Hint: Ensure that you test every vanishing partial reg-ression coefficient that is implied by the product decomposition (1.29).]

Bibliographical Notes for Chapter 2

The distinction between chains and forks in causal models was made by Simon (1953) and Reichenbach (1956) while the treatment of colliders (or common effect) can be traced back to the English economist Pigou (1911) (see Stigler 1999, pp. 36–41). In epidemiology, colliders came to be associated with "Selection bias" or "Berkson paradox" (Berkson 1946) while in artificial intelligence it came to be known as the "explaining away effect" (Kim and Pearl 1983). The rule of *d*-separation for determining conditional independence by graphs (Definition 2.4.1) was introduced in Pearl (1986) and formally proved in Verma and Pearl (1988) using the theory of graphoids (Pearl and Paz 1987). Gentle introductions to *d*-separation are available in Hayduk et al. (2003), Glymour and Greenland (2008), and Pearl (2000, pp. 335–337). Algorithms and software for detecting *d*-separation, as well as finding minimal separating sets are described in Tian et al. (1998), Kyono (2010), and Textor et al. (2011). The advantages of local over global model testing, are discussed in Pearl (2000, pp. 144–145) and further elaborated in Chen and Pearl (2014). Recent applications of *d*-separation include extrapolation across populations (Pearl and Bareinboim 2014), recovering from sampling selection bias (Bareinboim et al. 2014), and handling missing data (Mohan et al. 2013). A gentle introduction to *d*-separation and many of its applications can be found in *The Book of Why* (Pearl and Mackenzie 2018).

3

The Effects of Interventions

3.1 Interventions

The ultimate aim of many statistical studies is to predict the effects of interventions. When we collect data on factors associated with wildfires in the west, we are actually searching for something we can intervene upon in order to decrease wildfire frequency. When we perform a study on a new cancer drug, we are trying to identify how a patient's illness responds when we intervene upon it by medicating the patient. When we research the correlation between violent television and acts of aggression in children, we are trying to determine whether intervening to reduce children's access to violent television will reduce their aggressiveness.

As you have undoubtedly heard many times in statistics classes, "correlation is not causation." A mere association between two variables does not necessarily or even usually mean that one of those variables causes the other. (The famous example of this property is that an increase in ice cream sales is correlated with an increase in violent crime—not because ice cream causes crime, but because both ice cream sales and violent crime are more common in hot weather.) For this reason, the randomized controlled experiment is considered the golden standard of statistics. In a properly randomized controlled experiment, all factors that influence the outcome variable are either static, or vary at random, except for one—so any change in the outcome variable must be due to that one input variable.

Unfortunately, many questions do not lend themselves to randomized controlled experiments. We cannot control the weather, so we can't randomize the variables that affect wildfires. We could conceivably randomize the participants in a study about violent television, but it would be difficult to effectively control how much television each child watches, and nearly impossible to know whether we were controlling them effectively or not. Even randomized drug trials can run into problems when participants drop out, fail to take their medication, or misreport their usage.

In cases where randomized controlled experiments are not practical, researchers instead perform observational studies, in which they merely record data, rather than controlling it. The problem of such studies is that it is difficult to untangle the causal from the merely correlative. Our common sense tells us that intervening on ice cream sales is unlikely to have any effect on crime, but the facts are not always so clear. Consider, for instance, a recent

Causal Inference in Statistics: A Primer, First Edition. Judea Pearl, Madelyn Glymour, and Nicholas P. Jewell.

Companion Website: www.wiley.com/go/Pearl/Causality

University of Winnipeg study that showed that heavy text messaging in teens was correlated with "shallowness." Media outlets jumped on this as proof that texting makes teenagers more shallow. (Or, to use the language of intervention, that intervening to make teens text less would make them less shallow.) The study, however, proved nothing of the sort. It might be the case that shallowness makes teens more drawn to texting. It might be that both shallowness and heavy texting are caused by a common factor—a gene, perhaps—and that intervening on that variable, if possible, would decrease both.

The difference between intervening on a variable and conditioning on that variable should, hopefully, be obvious. When we intervene on a variable in a model, we fix its value. We *change* the system, and the values of other variables often change as a result. When we condition on a variable, we change nothing; we merely narrow our focus to the subset of cases in which the variable takes the value we are interested in. What changes, then, is our perception about the world, not the world itself.

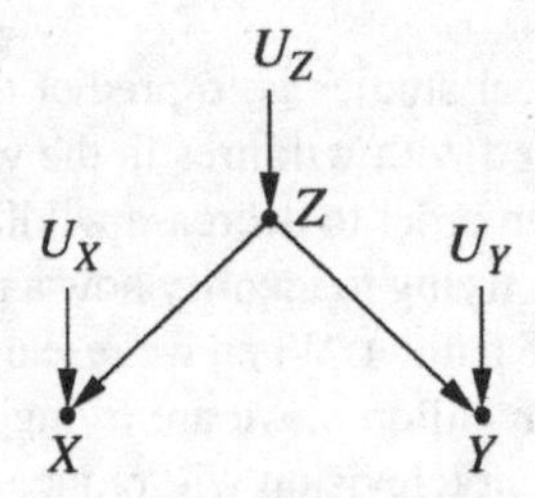

Figure 3.1 A graphical model representing the relationship between temperature (Z), ice cream sales (X), and crime rates (Y)

Consider, for instance, Figure 3.1 that shows a graphical model of our ice cream sales example, with X as ice cream sales, Y as crime rates, and Z as temperature. When we intervene to fix the value of a variable, we curtail the natural tendency of that variable to vary in response to other variables in nature. This amounts to performing a kind of surgery on the graphical model, removing all edges directed into that variable. If we were to intervene to make ice cream sales low (say, by shutting down all ice cream shops), we would have the graphical model shown in Figure 3.2. When we examine correlations in this new graph, we find that crime rates are, of course, totally independent of (i.e., uncorrelated with) ice cream sales since the latter is no longer associated with temperature (Z). In other words, even if we vary the level at which we hold X constant, that variation will not be transmitted to variable Y (crime rates). We see that intervening on a variable results in a totally different pattern of dependencies than conditioning on a variable. Moreover, the latter can be obtained

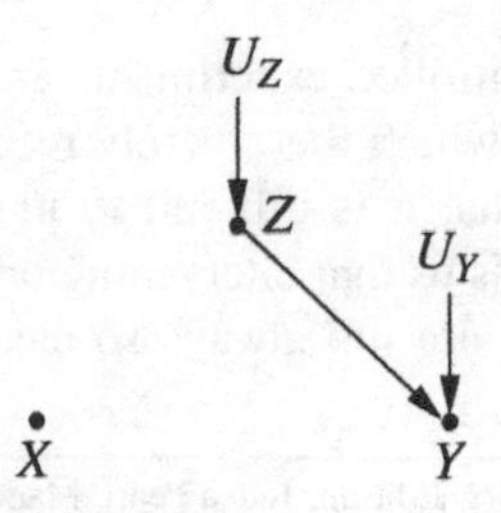

Figure 3.2 A graphical model representing an intervention on the model in Figure 3.1 that lowers ice cream sales

directly from the data set, using the procedures described in Part One, while the former varies depending on the structure of the causal graph. It is the graph that instructs us which arrow should be removed for any given intervention.

In notation, we distinguish between cases where a variable X takes a value x naturally and cases where we fix $X = x$ by denoting the latter $do(X = x)$. So $P(Y = y|X = x)$ is the probability that $Y = y$ conditional on finding $X = x$, while $P(Y = y|do(X = x))$ is the probability that $Y = y$ when we intervene to make $X = x$. In the distributional terminology, $P(Y = y|X = x)$ reflects the population distribution of Y among individuals whose X value is x. On the other hand, $P(Y = y|do(X = x))$ represents the population distribution of Y if *everyone in the population* had their X value fixed at x. We similarly write $P(Y = y|do(X = x), Z = z)$ to denote the conditional probability of $Y = y$, given $Z = z$, in the distribution created by the intervention $do(X = x)$.

Using *do*-expressions and graph surgery, we can begin to untangle the causal relationships from the correlative. In the rest of this chapter, we learn methods that can, astoundingly, tease out causal information from purely observational data, assuming of course that the graph constitutes a valid representation of reality. It is worth noting here that we are making a tacit assumption that the intervention has no "side effects," that is, that *assigning* the value x for the variable X for an individual does not alter subsequent variables in a direct way. For example, being "assigned" a drug might have a different effect on recovery than being forced to take the drug against one's religious objections. When side effects are present, they need to be specified explicitly in the model.

3.2 The Adjustment Formula

The ice cream example represents an extreme case in which the correlation between X and Y was totally spurious from a causal perspective, because there was no causal path from X to Y. Most real-life situations are not so clear-cut. To explore a more realistic situation, let us examine Figure 3.3, in which Y responds to both Z and X. Such a model could represent, for example, the first story we encountered for Simpson's paradox, where X stands for drug usage, Y stands for recovery, and Z stands for gender. To find out how effective the drug is in the population, we imagine a hypothetical intervention by which we administer the drug uniformly to the entire population and compare the recovery rate to what would obtain under the complementary intervention, where we prevent everyone from using the drug. Denoting the first intervention by $do(X = 1)$ and the second by $do(X = 0)$, our task is to estimate the difference

$$P(Y = 1|do(X = 1)) - P(Y = 1|do(X = 0)) \tag{3.1}$$

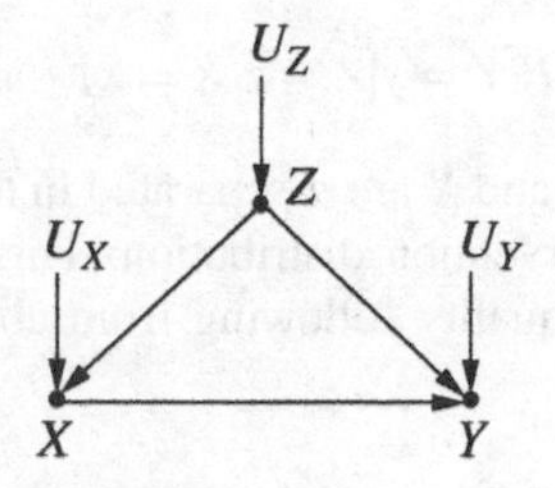

Figure 3.3 A graphical model representing the effects of a new drug, with Z representing gender, X standing for drug usage, and Y standing for recovery

which is known as the "causal effect difference," or "average causal effect" (ACE). In general, however, if X and Y can each take on more than one value, we would wish to predict the general causal effect $P(Y = y|do(X = x))$, where x and y are any two values that X and Y can take on. For example, x may be the dosage of the drug and y the patient's blood pressure.

We know from first principles that causal effects cannot be estimated from the data set itself without a causal story. That was the lesson of Simpson's paradox: The data itself was not sufficient even for determining whether the effect of the drug was positive or negative. But with the aid of the graph in Figure 3.3, we can compute the magnitude of the causal effect from the data. To do so, we simulate the intervention in the form of a graph surgery (Figure 3.4) just as we did in the ice cream example. The causal effect $P(Y = y|do(X = x))$ is equal to the conditional probability $P_m(Y = y|X = x)$ that prevails in the *manipulated* model of Figure 3.4. (This, of course, also resolves the question of whether the correct answer lies in the aggregated or the Z-specific table—when we determine the answer through an intervention, there's only one table to contend with.)

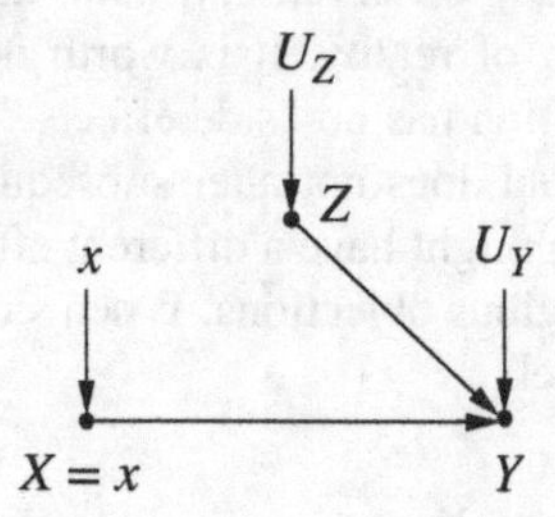

Figure 3.4 A modified graphical model representing an intervention on the model in Figure 3.3 that sets drug usage in the population, and results in the manipulated probability P_m

The key to computing the causal effect lies in the observation that P_m, the manipulated probability, shares two essential properties with P (the original probability function that prevails in the preintervention model of Figure 3.3). First, the marginal probability $P(Z = z)$ is invariant under the intervention, because the process determining Z is not affected by removing the arrow from Z to X. In our example, this means that the proportions of males and females remain the same, before and after the intervention. Second, the conditional probability $P(Y = y|Z = z, X = x)$ is invariant, because the process by which Y responds to X and Z, $Y = f(x, z, u_Y)$, remains the same, regardless of whether X changes spontaneously or by deliberate manipulation. We can therefore write two equations of invariance:

$$P_m(Y = y|Z = z, X = x) = P(Y = y|Z = z, X = x) \quad \text{and} \quad P_m(Z = z) = P(Z = z)$$

We can also use the fact that Z and X are d-separated in the modified model and are, therefore, independent under the intervention distribution. This tells us that $P_m(Z = z|X = x) = P_m(Z = z) = P(Z = z)$, the last equality following from above. Putting these considerations together, we have

$$P(Y = y|do(X = x))$$

$$= P_m(Y = y|X = x) \quad \text{(by definition)} \tag{3.2}$$

$$= \sum_z P_m(Y = y|X = x, Z = z)P_m(Z = z|X = x) \quad (3.3)$$

$$= \sum_z P_m(Y = y|X = x, Z = z)P_m(Z = z) \quad (3.4)$$

Equation (3.3) is obtained using the Law of Total Probability by conditioning on and summing over all values of $Z = z$ (as in Eq. (1.9)) while Eq. (3.4) makes use of the independence of Z and X in the modified model.

Finally, using the invariance relations, we obtain a formula for the causal effect, in terms of preintervention probabilities:

$$P(Y = y|do(X = x)) = \sum_z P(Y = y|X = x, Z = z)P(Z = z) \quad (3.5)$$

Equation (3.5) is called the *adjustment formula*, and as you can see, it computes the association between X and Y for each value z of Z, then averages over those values. This procedure is referred to as "adjusting for Z" or "controlling for Z."

This final expression—the right-hand side of Eq. (3.5)—can be estimated directly from the data, since it consists only of conditional probabilities, each of which can be computed by the filtering procedure described in Chapter 1. Note also that no adjustment is needed in a randomized controlled experiment since, in such a setting, the data are generated by a model which already possesses the structure of Figure 3.4, hence, $P_m = P$ regardless of any factors Z that affect Y. Our derivation of the adjustment formula (3.5) constitutes therefore a formal proof that randomization gives us the quantity we seek to estimate, namely $P(Y = y|do(X = x))$. In practice, investigators use adjustments in randomized experiments as well, for the purpose of minimizing sampling variations (Cox 1958).

To demonstrate the working of the adjustment formula, let us apply it numerically to Simpson's story, with $X = 1$ standing for the patient taking the drug, $Z = 1$ standing for the patient being male, and $Y = 1$ standing for the patient recovering. We have

$$P(Y = 1|do(X = 1)) = P(Y = 1|X = 1, Z = 1)P(Z = 1) + P(Y = 1|X = 1, Z = 0)P(Z = 0)$$

Substituting the figures given in Table 1.1 we obtain

$$P(Y = 1|do(X = 1)) = \frac{0.93(87 + 270)}{700} + \frac{0.73(263 + 80)}{700} = 0.832$$

while, similarly,

$$P(Y = 1|do(X = 0)) = \frac{0.87(87 + 270)}{700} + \frac{0.69(263 + 80)}{700} = 0.7818$$

Thus, comparing the effect of drug-taking ($X = 1$) to the effect of nontaking ($X = 0$), we obtain

$$ACE = P(Y = 1|do(X = 1)) - P(Y = 1|do(X = 0)) = 0.832 - 0.7818 = 0.0502$$

giving a clear positive advantage to drug-taking. A more informal interpretation of ACE here is that it is simply the difference in the fraction of the population that would recover if everyone took the drug compared to when no one takes the drug.

We see that the adjustment formula instructs us to condition on gender, find the benefit of the drug separately for males and females, and only then average the result using the percentage of males and females in the population. It also thus instructs us to ignore the aggregated population data $P(Y = 1|X = 1)$ and $P(Y = 1|X = 0)$, from which we might (falsely) conclude that the drug has a negative effect overall.

These simple examples might give readers the impression that whenever we face the dilemma of whether to condition on a third variable Z, the adjustment formula prefers the Z-specific analysis over the nonspecific analysis. But we know this is not so, recalling the blood pressure example of Simpson's paradox given in Table 1.2. There we argued that the more sensible method would be not to condition on blood pressure, but to examine the unconditional population table directly. How would the adjustment formula cope with situations like that?

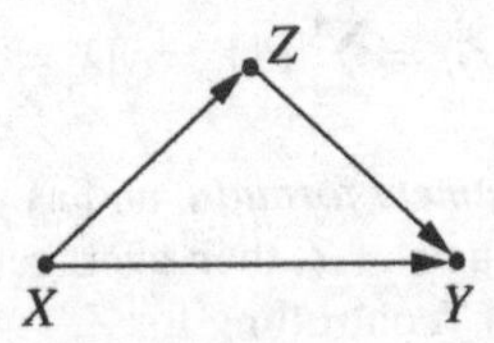

Figure 3.5 A graphical model representing the effects of a new drug, with X representing drug usage, Y representing recovery, and Z representing blood pressure (measured at the end of the study). Exogenous variables are not shown in the graph, implying that they are mutually independent

The graph in Figure 3.5 represents the causal story in the blood pressure example. It is the same as Figure 3.3, but with the arrow between X and Z reversed, reflecting the fact that the treatment has an effect on blood pressure and not the other way around. Let us try now to evaluate the causal effect $P(Y = 1|do(X = 1))$ associated with this model as we did with the gender example. First, we simulate an intervention and then examine the adjustment formula that emanates from the simulated intervention. In graphical models, an intervention is simulated by severing all arrows that enter the manipulated variable X. In our case, however, the graph of Figure 3.5 shows no arrow entering X, since X has no parents. This means that no surgery is required; the conditions under which data were obtained were such that treatment was assigned "as if randomized." If there was a factor that would make subjects prefer or reject treatment, such a factor should show up in the model; the absence of such a factor gives us the license to treat X as a randomized treatment.

Under such conditions, the intervention graph is equal to the original graph—no arrow need be removed—and the adjustment formula reduces to

$$P(Y = y|do(X = x)) = P(Y = y|X = x),$$

which can also be interpreted as a special case of the adjustment formula with the empty set being the element adjusted for. Obviously, if we were to adjust for blood pressure, we would obtain an incorrect assessment—one corresponding to a model in which blood pressure causes people to seek treatment.

3.2.1 To Adjust or not to Adjust?

We are now in a position to understand what variable, or set of variables, Z can legitimately be included in the adjustment formula. The intervention procedure, which led to the adjustment formula, dictates that Z should coincide with the parents of X, because it is the influence of these parents that we neutralize when we fix X by external manipulation. Denoting the parents of X by PA, we can therefore write a general adjustment formula and summarize it in a rule:

Rule 1 (The Causal Effect Rule) *Given a graph G in which a set of variables PA are designated as the parents of X, the causal effect of X on Y is given by*

$$P(Y = y|do(X = x)) = \sum_z P(Y = y|X = x, PA = z)P(PA = z) \tag{3.6}$$

where z ranges over all the combinations of values that the variables in PA can take.

If we multiply and divide the summand in (3.6) by the probability $P(X = x|PA = z)$, we get a more convenient form:

$$P(y|do(x)) = \sum_z \frac{P(X = x, Y = y, PA = z)}{P(X = x|PA = z)} \tag{3.7}$$

which explicitly displays the role played by the parents of X in predicting the results of interventions. The factor $P(X = x|PA = z)$ is known as the "propensity score" and the advantages of expressing $P(y|do(x))$ in this form will be discussed in Section 3.5.

We can appreciate now what role the causal graph plays in resolving Simpson's paradox, and, more generally, what aspects of the graph allow us to predict causal effects from purely statistical data. We need the graph in order to determine the identity of X's parents—the set of factors that, under nonexperimental conditions, would be sufficient for determining the value of X, or the probability of that value.

This result alone is astounding; using graphs and their underlying assumptions, we were able to identify causal relationships in purely observational data. But, from this discussion, readers may be tempted to conclude that the role of graphs is fairly limited; once we identify the parents of X, the rest of the graph can be discarded, and the causal effect can be evaluated mechanically from the adjustment formula. The next section shows that things may not be so simple. In most practical cases, the set of X's parents will contain unobserved variables that would prevent us from calculating the conditional probabilities in the adjustment formula. Luckily, as we will see in future sections, we can adjust for other variables in the model to substitute for the unmeasured elements of PA.

Study questions

Study questions 3.2.1

Referring to Study question 1.5.2 (Figure 1.10) and the parameters listed therein,

(a) Compute $P(y|do(x))$ for all values of x and y, by simulating the intervention do(x) on the model.
(b) Compute $P(y|do(x))$ for all values of x and y, using the adjustment formula (3.5)
(c) Compute the ACE

$$ACE = P(y_1|do(x_1)) - P(y_1|do(x_0))$$

and compare it to the Risk Difference

$$RD = P(y_1|x_1) - P(y_1|x_0)$$

What is the difference between ACE and the RD? What values of the parameters would minimize the difference?

(d) *Find a combination of parameters that exhibit Simpson's reversal (as in Study question 1.5.2(c)) and show explicitly that the overall causal effect of the drug is obtained from the disaggregated data.*

3.2.2 Multiple Interventions and the Truncated Product Rule

In deriving the adjustment formula, we assumed an intervention on a single variable, X, whose parents were disconnected, so as to simulate the absence of their influence after intervention. However, social and medical policies occasionally involve multiple interventions, such as those that dictate the value of several variables simultaneously, or those that control a variable over time. To represent multiple interventions, it is convenient to resort to the product decomposition that a graphical model imposes on joint distributions, as we have discussed in Section 1.5.2. According to the Rule of Product Decomposition, the preintervention distribution in the model of Figure 3.3 is given by the product

$$P(x, y, z) = P(z)P(x|z)P(y|x, z) \tag{3.8}$$

whereas the postintervention distribution, governed by the model of Figure 3.4 is given by the product

$$P(z, y|do(x)) = P_m(z)P_m(y|x, z) = P(z)P(y|x, z) \tag{3.9}$$

with the factor $P(x|z)$ purged from the product, since X becomes parentless as it is fixed at $X = x$. This coincides with the adjustment formula, because to evaluate $P(y|do(x))$ we need to marginalize (or sum) over z, which gives

$$P(y|do(x)) = \sum_z P(z)P(y|x, z)$$

in agreement with (3.5).

This consideration also allows us to generalize the adjustment formula to multiple interventions, that is, interventions that fix the values of a set of variables X to constants. We simply write down the product decomposition of the preintervention distribution, and strike out all factors that correspond to variables in the intervention set X. Formally, we write

$$P(x_1, x_2, \dots, x_n|do(x)) = \prod_i P(x_i|pa_i) \qquad \text{for all } i \text{ with } X_i \text{ not in } X.$$

This came to be known as the *truncated product formula* or *g*-formula. To illustrate, assume that we intervene on the model of Figure 2.9 and set X to x and Z_3 to z_3. The postintervention distribution of the other variables in the model will be

$$P(z_1, z_2, w, y|do(X = x, Z_3 = z_3)) = P(z_1)P(z_2)P(w|x)P(y|w, z_3, z_2)$$

where we have deleted the factors $P(x|z_1, z_3)$ and $P(z_3|z_1, z_2)$ from the product.

It is interesting to note that combining (3.8) and (3.9), we get a simple relation between the pre- and postintervention distributions:

$$P(z, y|do(x)) = \frac{P(x, y, z)}{P(x|z)} \tag{3.10}$$

It tells us that the conditional probability $P(x|z)$ is all we need to know in order to predict the effect of an intervention $do(x)$ from nonexperimental data governed by the distribution $P(x, y, z)$.

3.3 The Backdoor Criterion

In the previous section, we came to the conclusion that we should adjust for a variable's parents, when trying to determine its effect on another variable. But often, we know, or believe, that the variables have unmeasured parents that, though represented in the graph, may be inaccessible for measurement. In those cases, we need to find an alternative set of variables to adjust for.

This dilemma unlocks a deeper statistical question: Under what conditions does a causal story permit us to compute the causal effect of one variable on another, from data obtained by passive observations, with no interventions? Since we have decided to represent causal stories with graphs, the question becomes a graph-theoretical problem: Under what conditions is the structure of the causal graph sufficient for computing a causal effect from a given data set?

The answer to that question is long enough—and important enough—that we will spend the rest of the chapter addressing it. But one of the most important tools we use to determine whether we can compute a causal effect is a simple test called the *backdoor criterion*. Using it, we can determine, for any two variables X and Y in a causal model represented by a DAG, which set of variables Z in that model should be conditioned on when searching for the causal relationship between X and Y.

Definition 3.3.1 (The Backdoor Criterion) *Given an ordered pair of variables (X, Y) in a directed acyclic graph G, a set of variables Z satisfies the* backdoor criterion *relative to (X, Y) if no node in Z is a descendant of X, and Z blocks every path between X and Y that contains an arrow into X.*

If a set of variables Z satisfies the backdoor criterion for X and Y, then the causal effect of X on Y is given by the formula

$$P(Y = y|do(X = x)) = \sum_z P(Y = y|X = x, Z = z)P(Z = z)$$

just as when we adjust for PA. (Note that PA always satisfies the backdoor criterion.)

The logic behind the backdoor criterion is fairly straightforward. In general, we would like to condition on a set of nodes Z such that

1. We block all spurious paths between X and Y.
2. We leave all directed paths from X to Y unperturbed.
3. We create no new spurious paths.

When trying to find the causal effect of X on Y, we want the nodes we condition on to block any "backdoor" path in which one end has an arrow into X, because such paths may make X and Y dependent, but are obviously not transmitting causal influences from X, and if we do not block them, they will confound the effect that X has on Y. We condition on backdoor paths so as to fulfill our first requirement. However, we don't want to condition on any nodes that are descendants of X. Descendants of X would be affected by an intervention on X and might themselves affect Y; conditioning on them would block those pathways. Therefore, we don't condition on descendants of X so as to fulfill our second requirement. Finally, to comply with the third requirement, we should refrain from conditioning on any collider that would unblock a new path between X and Y. The requirement of excluding descendants of X also protects us from conditioning on children of intermediate nodes between X and Y (e.g., the collision node W in Figure 2.4.) Such conditioning would distort the passage of causal association between X and Y, similar to the way conditioning on their parents would.

To see what this means in practice, let's look at a concrete example, shown in Figure 3.6.

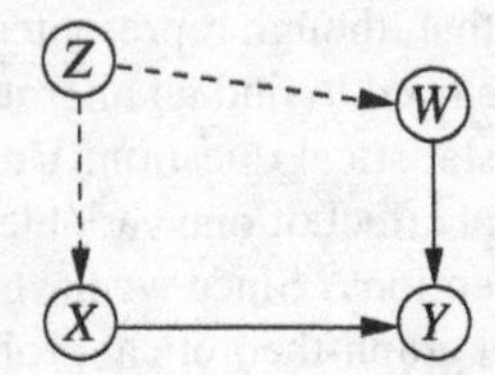

Figure 3.6 A graphical model representing the relationship between a new drug (X), recovery (Y), weight (W), and an unmeasured variable Z (socioeconomic status)

Here we are trying to gauge the effect of a drug (X) on recovery (Y). We have also measured weight (W), which has an effect on recovery. Further, we know that socioeconomic status (Z) affects both weight and the choice to receive treatment—but the study we are consulting did not record socioeconomic status.

Instead, we search for an observed variable that fits the backdoor criterion from X to Y. A brief examination of the graph shows that W, which is not a descendant of X, also blocks the backdoor path $X \leftarrow Z \rightarrow W \rightarrow Y$. Therefore, W meets the backdoor criterion. So long as the causal story conforms to the graph in Figure 3.6, adjusting for W will give us the causal effect of X on Y. Using the adjustment formula, we find

$$P(Y = y|do(X = x)) = \sum_{w} P(Y = y|X = x, W = w)P(W = w)$$

This sum can be estimated from our observational data, so long as W is observed.

With the help of the backdoor criterion, you can easily and algorithmically come to a conclusion about a pressing policy concern, even in complicated graphs. Consider the model in Figure 2.8, with its $X \rightarrow W$ edge replaced with forked path $X \leftarrow V \rightarrow W$, and assume again that we wish to evaluate the effect of X on Y. What variables should we condition on to obtain the correct effect? The question boils down to finding a set of variables that satisfy the backdoor criterion, but since there are no unblocked backdoor paths from X to Y, the answer is trivial: The empty set satisfies the criterion, hence no adjustment is needed. The answer is

$$P(y|do(x)) = P(y|x)$$

Suppose, however, that we were to adjust for W. Would we get the correct result for the effect of X on Y? Since W is a collider, conditioning on W would open the path $X \leftarrow V \rightarrow$

$W \leftarrow Z \leftarrow T \rightarrow Y$. This path is spurious since it lies outside the causal pathway from X to Y. Opening this path will create bias and yield an erroneous answer. This means that computing the association between X and Y for each value of W separately will not yield the correct effect of X on Y, and it might even give the wrong effect for each value of W.

How then do we compute the causal effect of X on Y for a specific value w of W? W may represent, for example, the level of a patient's preexisting pain (a consequence of a disease's severity, V), and we might be interested in assessing the effect of X on Y for only those patients who did not suffer any pain. Specifying the value of W amounts to conditioning on $W = w$, and this, as we have realized, opens a spurious path from X to Y by virtue of the fact that W is a collider.

The answer is that we still have the option of blocking that path using other variables. For example, if we condition on T, we would block the spurious path $X \leftarrow V \rightarrow W \leftarrow Z \leftarrow T \rightarrow Y$, even if W is part of the conditioning set. Thus to compute the w-specific causal effect, written $P(y|do(x), w)$, we adjust for T, and obtain

$$P(Y=y|do(X=x), W=w) = \sum_t P(Y=y|X=x, W=w, T=t)P(T=t|W=w) \tag{3.11}$$

Computing such W-specific causal effects is an essential step in examining *effect modification* or *moderation*, that is, the degree to which the causal effect of X on Y is modified by different values of W. Consider, again, the model in Figure 3.6, and suppose we wish to test whether the causal effect for units at level $W = w$ is the same as for units at level $W = w'$ (W may represent any pretreatment variable, such as age, sex, or ethnicity). This question calls for comparing two causal effects,

$$P(Y = y|do(X = x), W = w) \quad \text{and} \quad P(Y = y|do(X = x), W = w')$$

In the specific example of Figure 3.6, the answer is simple, because W satisfies the backdoor criterion. So, all we need to compare are the conditional probabilities $P(Y = y|X = x, W = w)$ and $P(Y = y|X = x, W = w')$; no summation is required. In the more general case, where W alone does not satisfy the backdoor criterion, yet a larger set, $T \cup W$, does, we need to adjust for members of T, which yields Eq. (3.11). We will return to this topic in Section 3.5.

From the examples seen thus far, readers may get the impression that one should refrain from adjusting for colliders. Such adjustment is sometimes unavoidable, as seen in Figure 3.7. Here, there are four backdoor paths from X to Y, all traversing variable Z, which is a collider on the path $X \leftarrow E \rightarrow Z \leftarrow A \rightarrow Y$. Conditioning on Z will unblock this path and will violate the backdoor criterion. To block all backdoor paths, we need to condition on one of the following sets: $\{E, Z\}$, $\{A, Z\}$, or $\{E, Z, A\}$. Each of these contains Z. We see, therefore, that Z, a collider, must be adjusted for in any set that yields an unbiased estimate of the effect of X on Y.

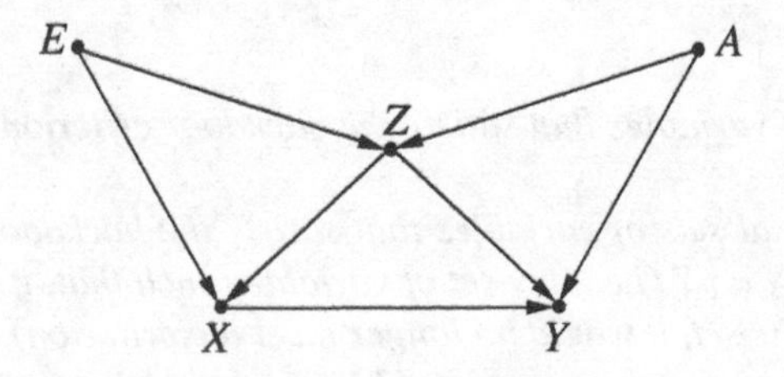

Figure 3.7 A graphical model in which the backdoor criterion requires that we condition on a collider (Z) in order to ascertain the effect of X on Y

The backdoor criterion has some further possible benefits. Consider the fact that $P(Y = y|do(X = x))$ is an empirical fact of nature, not a byproduct of our analysis. That means that any suitable variable or set of variables that we adjust on—whether it be PA or any other set that conforms to the backdoor criterion—must return the same result for $P(Y = y|do(X = x))$. In the case we looked at in Figure 3.6, this means that

$$\sum_w P(Y = y|X = x, W = w)P(W = w) = \sum_z P(Y = y|X = x, Z = z)P(Z = z)$$

This equality is useful in two ways. First, in the cases where we have multiple observed sets of variables suitable for adjustment (e.g., in Figure 3.6, if both W and Z had been observed), it provides us with a choice of which variables to adjust for. This could be useful for any number of practical reasons—perhaps one set of variables is more expensive to measure than the other, or more prone to human error, or simply has more variables and is therefore more difficult to calculate.

Second, the equality constitutes a testable constraint on the data when all the adjustment variables are observed, much like the rules of d-separation. If we are attempting to fit a model that leads to such an equality on a data set that violates it, we can discard that model.

Study questions

Study question 3.3.1

Consider the graph in Figure 3.8:

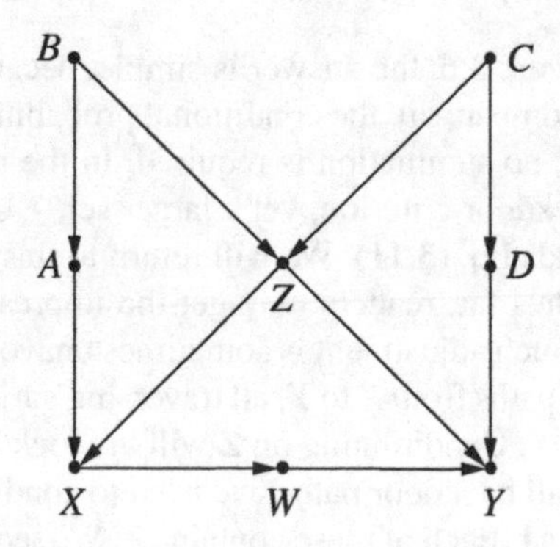

Figure 3.8 Causal graph used to illustrate the backdoor criterion in the following study questions

(a) List all of the sets of variables that satisfy the backdoor criterion to determine the causal effect of X on Y.

(b) List all of the minimal sets of variables that satisfy the backdoor criterion to determine the causal effect of X on Y (i.e., any set of variables such that, if you removed any one of the variables from the set, it would no longer meet the criterion).

(c) List all minimal sets of variables that need be measured in order to identify the effect of D on Y. Repeat, for the effect of {W, D} on Y.

Study question 3.3.2 (Lord's paradox)

At the beginning of the year, a boarding school offers its students a choice between two meal plans for the year: Plan A and Plan B. The students' weights are recorded at the beginning and the end of the year. To determine how each plan affects students' weight gain, the school hired two statisticians who, oddly, reached different conclusions. The first statistician calculated the difference between each student's weight in June (W_F) and in September (W_I) and found that the average weight gain in each plan was zero.

The second statistician divided the students into several subgroups, one for each initial weight, W_I. He finds that for each initial weight, the final weight for Plan B is higher than the final weight for Plan A.

So, the first statistician concluded that there was no effect of diet on weight gain and the second concluded there was.

Figure 3.9 illustrates data sets that can cause the two statisticians to reach conflicting conclusions. Statistician-1 examined the weight gain $W_F - W_I$, which, for each student, is represented by the shortest distance to the 45° line. Indeed, the average gain for each diet plan is zero; the two groups are each situated symmetrically relative to the zero-gain line, $W_F = W_I$. Statistician-2, on the other hand, compared the final weights of plan A students to those of plan B students who entered school with the same initial weight W_0 and, as the vertical line in the figure indicates, plan B students are situated above plan A students along this vertical line. The same will be the case for any other vertical line, regardless of W_0.

(a) Draw a causal graph representing the situation.
(b) Determine which statistician is correct.
(c) How is this example related to Simpson's paradox?

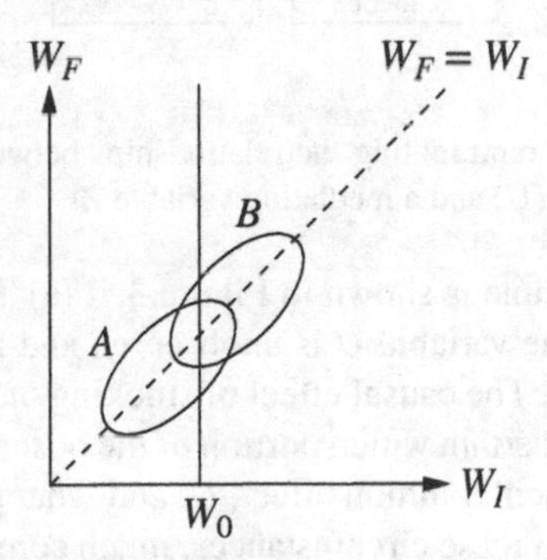

Figure 3.9 Scatter plot with students' initial weights on the x-axis and final weights on the y-axis. The vertical line indicates students whose initial weights are the same, and whose final weights are higher (on average) for plan B compared with plan A

Study questions 3.3.3

Revisit the lollipop story of Study question 1.2.4 and answer the following questions:

(a) Draw a graph that captures the story.
(b) Determine which variables must be adjusted for by applying the backdoor criterion.

(c) Write the adjustment formula for the effect of the drug on recovery.
(d) Repeat questions (a)–(c) assuming that the nurse gave lollipops a day after the study, still preferring patients who received treatment over those who received placebo.

3.4 The Front-Door Criterion

The backdoor criterion provides us with a simple method of identifying sets of covariates that should be adjusted for when we seek to estimate causal effects from nonexperimental data. It does not, however, exhaust *all* ways of estimating such effects. The *do*-operator can be applied to graphical patterns that do not satisfy the backdoor criterion to identify effects that on first sight seem to be beyond one's reach. One such pattern, called front-door, is discussed in this section.

Consider the century-old debate on the relation between smoking and lung cancer. In the years preceding 1970, the tobacco industry managed to prevent antismoking legislation by promoting the theory that the observed correlation between smoking and lung cancer could be explained by some sort of carcinogenic genotype that also induces an inborn craving for nicotine.

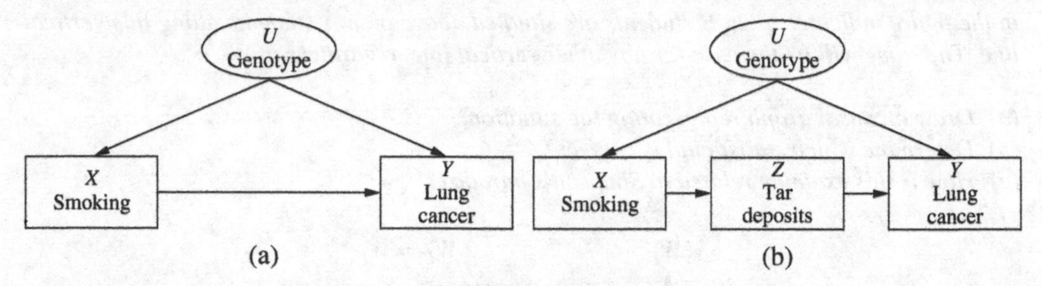

Figure 3.10 A graphical model representing the relationships between smoking (X) and lung cancer (Y), with unobserved confounder (U) and a mediating variable Z

A graph depicting this example is shown in Figure 3.10(a) This graph does not satisfy the backdoor condition because the variable U is unobserved and hence cannot be used to block the backdoor path from X to Y. The causal effect of smoking on lung cancer is not identifiable in this model; one can never ascertain which portion of the observed correlation between X and Y is spurious, attributable to their common effect, U, and what portion is genuinely causative. (We note, however, that even in these circumstances, much compelling work has been done to quantify how strong the (unobserved) associates between both U and X, and U and Y, must be in order to entirely explain the observed association between X and Y.)

However, we can go much further by considering the model in Figure 3.10(b), where an additional measurement is available: the amount of tar deposits in patients' lungs. This model does not satisfy the backdoor criterion, because there is still no variable capable of blocking the spurious path $X \leftarrow U \rightarrow Y$. We see, however, that the causal effect $P(Y = y|do(X = x))$ is nevertheless identifiable in this model, through two consecutive applications of the backdoor criterion.

How can the intermediate variable Z help us to assess the effect of X on Y? The answer is not at all trivial: as the following quantitative example shows, it may lead to heated debate.

Assume that a careful study was undertaken, in which the following factors were measured simultaneously on a randomly selected sample of 800,000 subjects considered to be at very high risk of cancer (because of environmental exposures such as smoking, asbestos, radon, and the like).

1. Whether the subject smoked
2. Amount of tar in the subject's lungs
3. Whether lung cancer has been detected in the patient.

The data from this study are presented in Table 3.1, where, for simplicity, all three variables are assumed to be binary. All numbers are given in thousands.

Table 3.1 A hypothetical data set of randomly selected samples showing the percentage of cancer cases for smokers and nonsmokers in each tar category (numbers in thousands)

	Tar 400		No tar 400		All subjects 800	
	Smokers	Nonsmokers	Smokers	Nonsmokers	Smokers	Nonsmokers
	380	*20*	*20*	*380*	*400*	*400*
No cancer	323 (85%)	1 (5%)	18 (90%)	38 (10%)	341 (85%)	39 (9.75%)
Cancer	57 (15%)	19 (95%)	2 (10%)	342 (90%)	59 (15%)	361 (90.25%)

Two opposing interpretations can be offered for these data. The tobacco industry argues that the table proves the beneficial effect of smoking. They point to the fact that only 15% of the smokers have developed lung cancer, compared to 90.25% of the nonsmokers. Moreover, within each of two subgroups, tar and no tar, smokers show a much lower percentage of cancer than nonsmokers. (These numbers are obviously contrary to empirical observations but well illustrate our point that observations are not to be trusted.)

However, the antismoking lobbyists argue that the table tells an entirely different story—that smoking would actually increase, not decrease, one's risk of lung cancer. Their argument goes as follows: If you choose to smoke, then your chances of building up tar deposits are 95%, compared to 5% if you choose not to smoke (380/400 vs 20/400). To evaluate the effect of tar deposits, we look separately at two groups, smokers and nonsmokers, as done in Table 3.2. All numbers are given in thousands.

Table 3.2 Reorganization of the data set of Table 3.1 showing the percentage of cancer cases in each smoking-tar category (numbers in thousands)

	Smokers 400		Nonsmokers 400		All subjects 800	
	Tar	No tar	Tar	No tar	Tar	No tar
	380	*20*	*20*	*380*	*400*	*400*
No cancer	323 (85%)	18 (90%)	1 (5%)	38 (10%)	324 (81%)	56 (19%)
Cancer	57 (15%)	2 (10%)	19 (95%)	342 (90%)	76 (19%)	344 (81%)

It appears that tar deposits have a harmful effect in both groups; in smokers it increases cancer rates from 10% to 15%, and in nonsmokers it increases cancer rates from 90% to 95%. Thus, regardless of whether I have a natural craving for nicotine, I should avoid the harmful effect of tar deposits, and no-smoking offers very effective means of avoiding them.

The graph of Figure 3.10(b) enables us to decide between these two groups of statisticians. First, we note that the effect of X on Z is identifiable, since there is no backdoor path from X to Z. Thus, we can immediately write

$$P(Z = z|do(X = x)) = P(Z = z|X = x) \tag{3.12}$$

Next we note that the effect of Z on Y is also identifiable, since the backdoor path from Z to Y, namely $Z \leftarrow X \leftarrow U \rightarrow Y$, can be blocked by conditioning on X. Thus, we can write

$$P(Y = y|do(Z = z)) = \sum_x P(Y = y|Z = z, X = x)P(X = x) \tag{3.13}$$

Both (3.12) and (3.13) are obtained through the adjustment formula, the first by conditioning on the null set, and the second by adjusting for X.

We are now going to chain together the two partial effects to obtain the overall effect of X on Y. The reasoning goes as follows: If nature chooses to assign Z the value z, then the probability of Y would be $P(Y = y|do(Z = z))$. But the probability that nature would choose to do that, given that we choose to set X at x, is $P(Z = z|do(X = x))$. Therefore, summing over all states z of Z, we have

$$P(Y = y|do(X = x)) = \sum_z P(Y = y|do(Z = z))P(Z = z|do(X = x)) \tag{3.14}$$

The terms on the right-hand side of (3.14) were evaluated in (3.12) and (3.13), and we can substitute them to obtain a *do*-free expression for $P(Y = y|do(X = x))$. We also distinguish between the x that appears in (3.12) and the one that appears in (3.13), the latter of which is merely an index of summation and might as well be denoted x'. The final expression we have is

$$P(Y = y|do(X = x)) = \sum_z \sum_{x'} P(Y = y|Z = z, X = x')P(X = x')P(Z = z|X = x) \tag{3.15}$$

Equation (3.15) is known as the *front-door formula*.

Applying this formula to the data in Table 3.1, we see that the tobacco industry was wrong; tar deposits have a harmful effect in that they make lung cancer more likely and smoking, by increasing tar deposits, increases the chances of causing this harm.

The data in Table 3.1 are obviously unrealistic and were deliberately crafted so as to surprise readers with counterintuitive conclusions that may emerge from naive analysis of observational data. In reality, we would expect observational studies to show positive correlation between smoking and lung cancer. The estimand of (3.15) could then be used for confirming and quantifying the harmful effect of smoking on cancer.

The preceding analysis can be generalized to structures where multiple paths lead from X to Y.

Definition 3.4.1 (Front-Door) *A set of variables Z is said to satisfy the front-door criterion relative to an ordered pair of variables (X, Y) if*

1. *Z intercepts all directed paths from X to Y.*
2. *There is no unblocked backdoor path from X to Z.*
3. *All backdoor paths from Z to Y are blocked by X.*

Theorem 3.4.1 (Front-Door Adjustment) *If Z satisfies the front-door criterion relative to (X, Y) and if $P(x, z) > 0$, then the causal effect of X on Y is identifiable and is given by the formula*

$$P(y|do(x)) = \sum_z P(z|x) \sum_{x'} P(y|x', z)P(x') \tag{3.16}$$

The conditions stated in Definition 3.4.1 are overly conservative; some of the paths excluded by conditions (2) and (3) can actually be allowed provided they are blocked by some variables. There is a powerful symbolic machinery, called the *do-calculus*, that allows analysis of such intricate structures. In fact, the *do*-calculus uncovers *all* causal effects that can be identified from a given graph. Unfortunately, it is beyond the scope of this book (see Tian and Pearl 2002, Shpitser and Pearl 2008, Pearl 2009, and Bareinboim and Pearl 2012 for details). But the combination of the adjustment formula, the backdoor criterion, and the front-door criterion covers numerous scenarios. It proves the enormous, even revelatory, power that causal graphs have in not merely representing, but actually discovering causal information.

Study questions

Study question 3.4.1

Assume that in Figure 3.8, only X, Y, and one additional variable can be measured. Which variable would allow the identification of the effect of X on Y? What would that effect be?

Study question 3.4.2

I went to a pharmacy to buy a certain drug, and I found that it was available in two different bottles: one priced at $1, the other at $10. I asked the druggist, "What's the difference?" and he told me, "The $10 bottle is fresh, whereas the $1 bottle one has been on the shelf for 3 years. But, you know, data shows that the percentage of recovery is much higher among those who bought the cheap stuff. Amazing isn't it?" I asked if the aged drug was ever tested. He said, "Yes, and this is even more amazing; 95% of the aged drug and only 5% of the fresh drug has lost the active ingredient, yet the percentage of recovery among those who got bad bottles, with none of the active ingredient, is still much higher than among those who got good bottles, with the active ingredient."

Before ordering a cheap bottle, it occurred to me to have a good look at the data. The data were, for each previous customer, the type of bottle purchased (aged or fresh), the concentration of the active ingredient in the bottle (high or low), and whether the customer recovered from the illness. The data perfectly confirmed the druggist's story. However, after making some additional calculations, I decided to buy the expensive bottle after all; even without testing its content, I could determine that a fresh bottle would offer the average patient a greater chance of recovery.

Based on two very reasonable assumptions, the data show clearly that the fresh drug is more effective. The assumptions are as follows:

(i) Customers had no information about the chemical content (high or low) of the specific bottle of the drug that they were buying; their choices were influenced by price and shelf-age alone.

(ii) The effect of the drug on any given individual depends only on its chemical content, not on its shelf age (fresh or aged).

(a) Determine the relevant variables for the problem, and describe this scenario in a causal graph.

(b) Construct a data set compatible with the story and the decision to buy the expensive bottle.

(c) Determine the effect of choosing the fresh versus the aged drug by using assumptions (i) and (ii), and the data given in (b).

3.5 Conditional Interventions and Covariate-Specific Effects

The interventions considered thus far have been limited to actions that merely force a variable or a group of variables X to take on some specified value x. In general, interventions may involve dynamic policies in which a variable X is made to respond in a specified way to some set Z of other variables—say, through a functional relationship $x = g(z)$ or through a stochastic relationship, whereby X is set to x with probability $P^*(x|z)$. For example, suppose a doctor decides to administer a drug only to patients whose temperature Z exceeds a certain level, $Z = z$. In this case, the action will be *conditional* upon the value of Z and can be written $do(X = g(Z))$, where $g(Z)$ is equal to one when $Z > z$ and zero otherwise (where $X = 0$ represents no drug). Since Z is a random variable, the value of X chosen by the action will similarly be a random variable, tracking variations in Z. The result of implementing such a policy is a probability distribution written $P(Y = y|do(X = g(Z)))$, which depends only on the function g and the set Z of variables that drive X.

In order to estimate the effect of such a policy, let us take a closer look at another concept, the "z-specific effect" of X, which we encountered briefly in Section 3.3 (Eq. (3.11)). This effect, written $P(Y = y|do(X = x), Z = z)$, measures the distribution of Y in a subset of the population for which Z achieves the value z after the intervention. For example, we may be interested in how a treatment affects a specific age group, $Z = z$, or people with a specific feature, $Z = z$, which may be measured after the treatment.

The z-specific effect can be identified by a procedure similar to the backdoor adjustment. The reasoning goes as follows: When we aim to estimate $P(Y = y|do(X = x))$, an adjustment for a set S is justified if S blocks all backdoor paths from X to Y. Now that we wish to identify $P(Y = y|do(X = x), Z = z)$, we need to ensure that those paths remain blocked when we add one more variable, Z, to the conditioning set. This yields a simple criterion for the identification of the z-specific effect:

Rule 2 *The z-specific effect* $P(Y = y|do(X = x), Z = z)$ *is identified whenever we can measure a set S of variables such that* $S \cup Z$ *satisfies the backdoor criterion. Moreover, the z-specific* *effect is given by the following adjustment formula*

$$
\begin{aligned}
&P(Y = y|do(X = x), Z = z) \\
&\quad = \sum_s P(Y = y|X = x, S = s, Z = z)P(S = s|Z = z)
\end{aligned}
$$

This modified adjustment formula is similar to Eq. (3.5) with two exceptions. First, the adjustment set is $S \cup Z$, not just S and, second, the summation goes only over S, not including Z. The $\cup$ symbol in the expression $S \cup Z$ stands for set addition (or union), which means that, if Z is a subset of S, we have $S \cup Z = S$, and S alone need satisfy the backdoor criterion.

Note that the identifiability criterion for z-specific effects is somewhat stricter than that for nonspecific effect. Adding Z to the conditioning set might create dependencies that would prevent the blocking of all backdoor paths. A simple example occurs when Z is a collider; conditioning on Z will create a new dependency between Z's parents and may thus violate the backdoor requirement.

We are now ready to tackle our original task of estimating conditional interventions. Suppose a policy maker contemplates an age-dependent policy whereby an amount x of drug is to be administered to patients, depending on their age Z. We write it as $do(X = g(Z))$. To find out the distribution of outcome Y that results from this policy, we seek to estimate $P(Y = y|do(X = g(Z)))$.

We now show that identifying the effect of such policies is equivalent to identifying the expression for the z-specific effect $P(Y = y|do(X = x), Z = z)$.

To compute $P(Y = y|do(X = g(Z)))$, we condition on $Z = z$ and write

$$
\begin{aligned}
&P(Y = y|do(X = g(Z))) \\
&\quad = \sum_z P(Y = y|do(X = g(Z)), Z = z)P(Z = z|do(X = g(Z))) \\
&\quad = \sum_z P(Y = y|do(X = g(z)), Z = z)P(Z = z) \qquad (3.17)
\end{aligned}
$$

The equality

$$P(Z = z|do(X = g(Z))) = P(Z = z)$$

stems, of course, from the fact that Z occurs before X; hence, any control exerted on X can have no effect on the distribution of Z. Equation (3.17) can also be written as

$$\sum_z P(Y = y|do(X = x), Z = z)|_{x=g(z)} P(Z = z)$$

which tells us that the causal effect of a conditional policy $do(X = g(Z))$ can be evaluated directly from the expression of $P(Y = y|do(X = x), Z = z)$ simply by substituting $g(z)$ for x and taking the expectation over Z (using the observed distribution $P(Z = z)$).

Study question 3.5.1

Consider the causal model of Figure 3.8.

(a) Find an expression for the c-specific effect of X on Y.

(b) *Identify a set of four variables that need to be measured in order to estimate the z-specific effect of X on Y, and find an expression for the size of that effect.*
(c) *Using your answer to part (b), determine the expected value of Y under a Z-dependent strategy, where X is set to 0 when Z is smaller or equal to 2 and X is set to 1 when Z is larger than 2. (Assume Z takes on integer values from 1 to 5.)*

3.6 Inverse Probability Weighting

By now, the astute reader may have noticed a problem with our intervention procedures. The backdoor and front-door criteria tell us whether it is possible to predict the results of hypothetical interventions from data obtained in an observational study. Moreover, they tell us that we can make this prediction without simulating the intervention and without even thinking about it. All we need to do is identify a set Z of covariates satisfying one of the criteria, plug this set into the adjustment formula, and we're done: the resulting expression is guaranteed to provide a valid prediction of how the intervention will affect the outcome.

This is lovely in theory, but in practice, adjusting for Z may prove problematic. It entails looking at each value or combination of values of Z separately, estimating the conditional probability of Y given X in that stratum and then averaging the results. As the number of strata increases, adjusting for Z will encounter both computational and estimational difficulties. Since the set Z can be comprised of dozens of variables, each spanning dozens of discrete values, the summation required by the adjustment formula may be formidable, and the number of data samples falling within each $Z = z$ cell may be too small to provide reliable estimates of the conditional probabilities involved.

All of our work in this chapter has not been for naught, however. The adjustment procedure is straightforward, and, therefore, easy to use in the explanation of intervention criteria. But there is another, more subtle procedure that overcomes the practical difficulties of adjustment.

In this section, we discuss one way of circumventing this problem, provided only that we can obtain a reliable estimate of the function $g(x, z) = P(X = x|Z = z)$, often called the "propensity score," for each x and z. Such an estimate can be obtained by fitting the parameters of a flexible function $g(x, z)$ to the data at hand, in much the same way that we fitted the coefficients of a linear regression function, so as to minimize the mean square error with respect to a set of samples (Figure 1.4). The method used will depend on the nature of the random variable X, whether it is continuous, discrete, or binary, for example.

Assuming that the function $P(X = x|Z = z)$ is available to us, we can use it to generate artificial samples that act as though they were drawn from the postintervention probability P_m, rather than $P(x, y, z)$. Once we obtain such fictitious samples, we can evaluate $P(Y = y|do(x))$ by simply counting the frequency of the event $Y = y$, for each stratum $X = x$ in the sample. In this way, we skip the labor associated with summing over all strata $Z = z$; we essentially let nature do the summation for us.

The idea of estimating probabilities using fictitious samples is not new to us; it was used all along, though implicitly, whenever we estimated conditional probabilities from finite samples.

In Chapter 1, we characterized conditioning as a process of filtering—that is, ignoring all cases for which the condition $X = x$ does not hold, and normalizing the surviving cases, so that their total probabilities would add up to one. The net result of this operation is that the probability of each surviving case is boosted by a factor $1/P(X = x)$. This can be seen directly from Bayes' rule, which tells us that

$$P(Y = y, Z = z|X = x) = \frac{P(Y = y, Z = z, X = x)}{P(X = x)}$$

In other words, to find the probability of each row in the surviving table, we multiply the unconditional probability, $P(Y = y, Z = z, X = x)$ by the constant $1/P(X = x)$.

Let us now examine the population created by the $do(X = x)$ operation and ask how the probability of each case changes as a result of this operation. The answer is given to us by the adjustment formula, which reads

$$P(y|do(x)) = \sum_z P(Y = y|X = x, Z = z)P(Z = z)$$

Multiplying and dividing the expression inside the sum by the propensity score $P(X = x|Z = z)$, we get

$$P(y|do(x)) = \sum_z \frac{P(Y = y|X = x, Z = z)P(X = x|Z = z)P(Z = z)}{P(X = x|Z = z)}$$

Upon realizing the numerator is none other but the pretreatment distribution of (X, Y, Z), we can write

$$P(y|do(x)) = \sum_z \frac{P(Y = y, X = x, Z = z)}{P(X = x|Z = z)}$$

and the answer becomes clear: each case $(Y = y, X = x, Z = z)$ in the population should boost its probability by a factor equal to $1/P(X = x|Z = z)$. (Hence the name "inverse probability weighting.")

This provides us with a simple procedure of estimating $P(Y = y|do(X = x))$ when we have finite samples. If we weigh each available sample by a factor $= 1/P(X = x|Z = z)$, we can then treat the reweighted samples as if they were generated from P_m, not P, and proceed to estimate $P(Y = y|do(x))$ accordingly.

This is best demonstrated in an example.

Table 3.3 returns to our Simpson's paradox example of the drug that seems to help men and women but to hurt the general population. We'll use the same data we used before but presented

Table 3.3 Joint probability distribution $P(X, Y, Z)$ for the drug-gender-recovery story of Chapter 1 (Table 1.1)

X	Y	Z	% of population
Yes	Yes	Male	0.116
Yes	Yes	Female	0.274
Yes	No	Male	0.009
Yes	No	Female	0.101
No	Yes	Male	0.334
No	Yes	Female	0.079
No	No	Male	0.051
No	No	Female	0.036

Table 3.4 Conditional probability distribution $P(Y, Z|X)$ for drug users ($X = yes$) in the population of Table 3.3

X	Y	Z	% of population
Yes	Yes	Male	0.231
Yes	Yes	Female	0.549
Yes	No	Male	0.017
Yes	No	Female	0.203

this time as a weighted table. In this case, X represents whether or not the patient took the drug, Y represents whether the patient recovered, and Z represents the patient's gender.

If we condition on $X = Yes$, we get the data set shown in Table 3.4, which was formed in two steps. First, all rows with $X = No$ were excluded. Second, the weights given to the remaining rows were "renormalized," that is, multiplied by a constant so as to make them sum to one. This constant, according to Bayes' rule, is $1/P(X = yes)$, and $P(X = yes)$, in our example, is the combined weight of the first four rows of Table 3.3, which amounts to

$$P(X = yes) = 0.116 + 0.274 + 0.01 + 0.101 = 0.501$$

The result is the weight distribution in the four rows of Table 3.4; the weight of each row has been boosted by a factor $1/0.501 = 2.00$.

Let us now examine the population created by the $do(X = yes)$ operation, representing a deliberate decision to administer the drug to the same population.

To calculate the distribution of weights in this population, we need to compute the factor $P(X = yes|Z = z)$ for each z, which, according to Table 3.3, is given by

$$P(X = yes|Z = Male) = \frac{(0.116 + 0.01)}{(0.116 + 0.01 + 0.334 + 0.051)} = 0.247$$

$$P(X = yes|Z = Female) = \frac{(0.274 + 0.101)}{(0.274 + 0.101 + 0.079 + 0.036)} = 0.765$$

Multiplying the gender-matching rows by $1/0.247$ and $1/0.765$, respectively, we obtain Table 3.5, which represents the postintervention distribution of the population of Table 3.3. The probability of recovery in this distribution can now be computed directly from the data, by summing the first two rows:

$$P(Y = yes|do(X = yes)) = 0.470 + 0.358 = 0.828$$

Table 3.5 Probability distribution for the population of Table 3.3 under the intervention $do(X = Yes)$, determined via the inverse probability method

X	Y	Z	% of population
Yes	Yes	Male	0.470
Yes	Yes	Female	0.358
Yes	No	Male	0.040
Yes	No	Female	0.132

Three points are worth noting about this procedure. First, the redistribution of weight is no longer proportional but quite discriminatory. Row #1, for instance, boosted its weight from 0.116 to 0.476, a factor of 4.1, whereas Row #2 is boosted from 0.274 to 0.357, a factor of only 1.3. This redistribution renders X independent of Z, as in a randomized trial (Figure 3.4).

Second, an astute reader would notice that in this example no computational savings were realized; to estimate $P(Y = yes|do(X = yes))$ we still needed to sum over all values of Z, males and females. Indeed, the savings become significant when the number of Z values is in the thousands or millions, and the sample size is in the hundreds. In such cases, the number of Z values that the inverse probability method would encounter is equal to the number of samples available, not to the number of possible Z values, which is prohibitive.

Finally, an important word of caution. The method of inverse probability weighting is only valid when the set Z entering the factor $1/P(X = x|Z = z)$ satisfies the backdoor criterion. Lacking this assurance, the method may actually introduce more bias than the one obtained through naive conditioning, which produces Table 3.4 and the absurdities of Simpson's paradox.

Up to this point, and in the following, we focus on unbiased estimation of causal effects. In other words, we focus on estimates that will converge to the true causal effects as the number of samples increases indefinitely.

This is obviously important, but it is not the *only* issue relevant to estimation. In addition, we must also address *precision*. Precision refers to the variability of our causal estimates if the number of samples is finite, and, in particular, how much our estimate would vary from experiment to experiment. Clearly, all other things being equal, we prefer estimation procedures with high precision in addition to their possessing little or no bias. Practically, high-precision estimates lead to shorter confidence intervals that quantify our level of certainty as to how our sample estimates describe the causal effect of interest. Most of our discussion does not address the "best," or most precise, way to estimate relevant causal means and effects but focuses on whether it is possible to estimate such quantities from observed data distributions, when the number of samples goes to infinity.

For example, suppose we wish to estimate the causal effect of X on Y (in a causal graph as above), where X and Y both reflect continuous variables. Suppose the effect of Z is to make both high and low values of X most commonly observed, with values close to the middle of the range of X much less common. Then, inverse probability weighting down-weights the extreme values of X on both ends of its range (since these are observed most frequently due to Z) and essentially focuses entirely on the "middle" values of X. If we then use a regression model to estimate the causal effect of X on Y (see Section 3.8, for example) using the reweighted observations to account for the role of Z, the resulting estimates will be very imprecise. In such cases, we usually seek for alternative estimation strategies that are more precise. While we do not pursue these alternatives in this book, it is important to emphasize that, in addition to seeing that causal effects can be identified from the data, we must also devise effective strategies of using finite data to estimate effect sizes.

3.7 Mediation

Often, when one variable causes another, it does so both directly and indirectly, through a set of mediating variables. For instance, in our blood pressure/treatment/recovery example of Simpson's paradox, treatment is both a direct (negative) cause of recovery, and an indirect (positive) cause, through the mediator of blood pressure—treatment decreases blood pressure, which increases recovery. In many cases, it is useful to know how much of variable X's effect on variable Y is direct and how much is mediated. In practice, however, separating these two avenues of causation has proved difficult.

Suppose, for example, we want to know whether and to what degree a company discriminates by gender (X) in its hiring practices (Y). Such discrimination would constitute a direct effect of gender on hiring, which is illegal in many cases. However, gender also affects hiring practices in other ways; often, for instance, women are more or less likely to go into a particular field than men, or to have achieved advanced degrees in that field. So gender may also have an indirect effect on hiring through the mediating variable of qualifications (Z).

In order to find the direct effect of gender on hiring, we need to somehow hold qualifications steady, and measure the remaining relationship between gender and hiring; with qualifications unchanging, any change in hiring would have to be due to gender alone. Traditionally, this has been done by conditioning on the mediating variable. So if *P(Hired|Female, Highly Qualified)* is different from *P(Hired|Male, Highly Qualified)*, the reasoning goes, then there is a direct effect of gender on hiring.

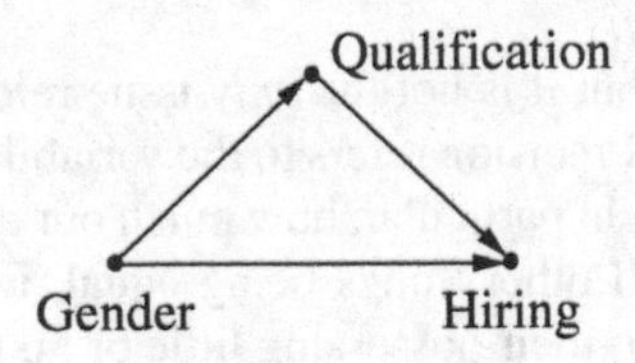

Figure 3.11 A graphical model representing the relationship between gender, qualifications, and hiring

In the example in Figure 3.11, this is correct. But consider what happens if there are confounders of the mediating variable and the outcome variable. For instance, income: People from higher income backgrounds are more likely to have gone to college and more likely to have connections that would help them get hired.

Now, if we condition on qualifications, we are conditioning on a collider. So if we don't condition on qualifications, indirect dependence can pass from gender to hiring through the path *Gender* → *Qualifications* → *Hiring*. But if we do condition on qualifications, indirect dependence can pass from gender to hiring through the path *Gender* → *Qualifications* ← *Income* → *Hiring*. (To understand the problem intuitively, note that by conditioning on qualification, we will be comparing men and women at different levels of income, because income must change to keep qualification constant.) No matter how you look at it, we're not getting the true direct effect of gender on hiring. Traditionally, therefore, statistics has had to abandon a huge class of potential mediation problems, where the concept of "direct effect" could not be defined, let alone estimated.

Luckily, we now have a conceptual way of holding the mediating variable steady without conditioning on it: We can intervene on it. If, instead of conditioning, we fix the qualifications, the arrow between gender and qualifications (and the one between income and qualifications) disappears, and no spurious dependence can pass through it. (Of course, it would be impossible for us to literally change the qualifications of applicants, but recall, this is a theoretical intervention of the kind discussed in the previous section, accomplished by choosing a proper adjustment.) So for any three variables X, Y, and Z, where Z is a mediator between X and Y, the *controlled direct effect* (CDE) on Y of changing the value of X from x' to x is defined as

$$CDE = P(Y = y|do(X = x), do(Z = z)) - P(Y = y|do(X = x'), do(Z = z)) \tag{3.18}$$

The obvious advantage of this definition over the one based on conditioning is its generality; it captures the intent of "keeping Z constant" even in cases where the $Z \rightarrow Y$ relationship is confounded (the same goes for the $X \rightarrow Z$ and $X \rightarrow Y$ relationships). Practically, this definition assures us that in any case where the intervened probabilities are identifiable from the observed probabilities, we can estimate the direct effect of X on Y. Note that the direct effect may differ for different values of Z; for instance, it may be that hiring practices discriminate against women in jobs with high qualification requirements, but they discriminate against men in jobs with low qualifications. Therefore, to get the full picture of the direct effect, we'll have to perform the calculation for every relevant value z of Z. (In linear models, this will not be necessary; for more information, see Section 3.8.)

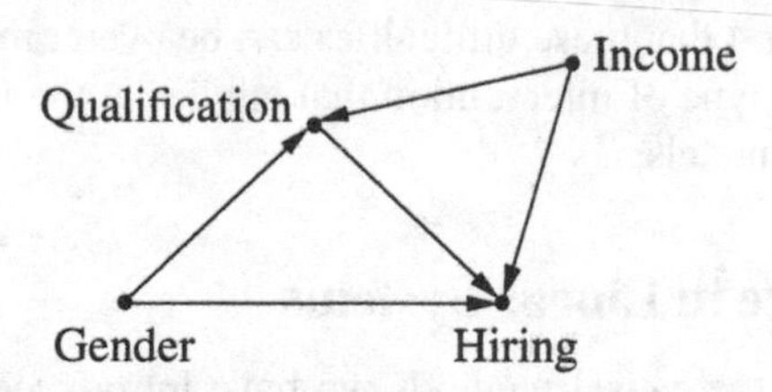

Figure 3.12 A graphical model showing qualification (Z) as a mediator between gender (X) and hiring (Y), and income (I) as a confounder between qualification and hiring.

How do we estimate the direct effect when its expression contains two *do*-operators? The technique is more or less the same as the one employed in Section 3.2, where we dealt with a single *do*-operator by adjustment. In our example of Figure 3.12, we first notice that there is no backdoor path from X to Y in the model, hence we can replace $do(x)$ with simply conditioning on x (this essentially amounts to adjusting for all confounders). This results in

$$P(Y = y|X = x, do(Z = z)) - P(Y = y|X = x', do(Z = z))$$

Next, we attempt to remove the $do(z)$ term and notice that two backdoor paths exist from Z to Y, one through X and one through I. The first is blocked (since X is conditioned on) and the second can be blocked if we adjust for I. This gives

$$\sum_i [P(Y = y|X = x, Z = z, I = i) - P(Y = y|X = x', Z = z, I = i)]P(I = i)$$

The last formula is *do*-free, which means it can be estimated from nonexperimental data.

In general, the CDE of X on Y, mediated by Z, is identifiable if the following two properties hold:

1. There exists a set S_1 of variables that blocks all backdoor paths from Z to Y.
2. There exists a set S_2 of variables that blocks all backdoor paths from X to Y, after deleting all arrows entering Z.

If these two properties hold in a model M, then we can determine $P(Y = y|do(X = x), do(Z = z))$ from the data set by adjusting for the appropriate variables, and estimating the conditional probabilities that ensue. Note that condition 2 is not necessary in randomized trials, because randomizing X renders X parentless. The same is true in cases where X is judged to be exogenous (i.e., "as if" randomized), as in the aforementioned gender discrimination example.

It is even trickier to determine the indirect effect than the direct effect, because there is simply no way to condition away the direct effect of X on Y. It's easy enough to find the total effect and the direct effect, so some may argue that the indirect effect should just be the difference between those two. This may be true in linear systems, but in nonlinear systems, differences don't mean much; the change in Y might, for instance, depend on some interaction between X and Z—if, as we posited above, women are discriminated against in high-qualification jobs and men in low-qualification jobs, subtracting the direct effect from the total effect would tell us very little about the effect of gender on hiring as mediated by qualifications. Clearly, we need a definition of indirect effect that does not depend on the total or direct effects.

We will show in Chapter 4 that these difficulties can be overcome through the use of *counterfactuals*, a more refined type of intervention that applies at the individual level and can be computed from structural models.

3.8 Causal Inference in Linear Systems

One of the advantages of the causal methods we have introduced in this book is that they work regardless of the type of equations that make up the model in question. d-separation and the backdoor criterion make no assumptions about the form of the relationship between two variables—only that the relationship exists.

However, showcasing and explaining causal methods from a nonparametric standpoint has limited our ability to present the full power of these methods as they play out in linear systems—the arena where traditional causal analysis has primarily been conducted in the social and behavioral sciences. This is unfortunate, as many statisticians work extensively in linear systems, and nearly all statisticians are very familiar with them.

In this section, we examine in depth what causal assumptions and implications look like in systems of linear equations and how graphical methods can help us answer causal questions posed in those systems. This will serve as both a reinforcement of the methods we applied in nonparametric models and as a useful aid for those hoping to apply causal inference specifically in the context of linear systems.

For instance, we might want to know the effect of birth control use on blood pressure after adjusting for confounders; the total effect of an after-school study program on test scores; the direct effect, unmediated by other variables, of the program on test scores; or the effect of enrollment in an optional work training program on future earnings, when enrollment and earnings are confounded by a common cause (e.g., motivation). Such questions, invoking continuous variables, have traditionally been formulated as linear equation models with only minor attention to the unique causal character of those equations; we make this character unambiguous.

In all models used in this section, we make the strong assumption that the relationships between variables are linear, and that all error terms have Gaussian (or "normal") distributions (in some cases, we only need to assume symmetric distributions). This assumption provides an enormous simplification of the procedure needed for causal analysis. We are all familiar with the bell-shaped curve that characterizes the normal distribution of one variable. The reason it is so popular in statistics is that it occurs so frequently in nature whenever a phenomenon is a byproduct of many noisy microprocesses that add up to produce macroscopic measurements such as height, weight, income, or mortality. Our interest in the normal distribution, however, stems primarily from the way several normally distributed variables combine to shape their joint distribution. The assumption of normality gives rise to four properties that are of enormous use when working with linear systems:

1. Efficient representation
2. Substitutability of expectations for probabilities
3. Linearity of expectations
4. Invariance of regression coefficients.

Starting with two normal variables, X and Y, we know that their joint density forms a three-dimensional cusp (like a mountain rising above the X–Y plane) and that the planes of equal height on that cusp are ellipses like those shown in Figure 1.2. Each such ellipse is characterized by five parameters: $\mu_X, \mu_Y, \sigma_X, \sigma_Y$, and ρ_{XY}, as defined in Sections 1.3.8 and 1.3.9. The parameters μ_X and μ_Y specify the location (or the center of gravity) of the ellipse in the X–Y plane, the standard deviations σ_X and σ_Y specify the spread of the ellipse along the X and Y dimensions, respectively, and the correlation coefficient ρ_{XY} specifies its orientation. In three dimensions, the best way to depict the joint distribution is to imagine an oval football sus-pended in the X–Y–Z space (Figure 1.2); every plane of constant Z would then cut the football in a two-dimensional ellipse like the ones shown in Figure 1.1.

As we go to higher dimensions, and consider a set of N normally distributed variables $X_1, X_2, \ldots, X_N$, we need not concern ourselves with additional parameters; it is sufficient to specify those that characterize the $N(N-1)/2$ pairs of variables, (X_i, X_j). In other words, the joint density of $(X_1, X_2, \ldots, X_N)$ is fully specified once we specify the bivariate density of (X_i, X_j), with i and j $(i \neq j)$ ranging from 1 to N. This is an enormously useful property, as it offers an extremely parsimonious way of specifying the N-variable joint distribution. Moreover, since the joint distribution of each pair is specified by five parameters, we conclude that the joint distribution requires at most $5 \times N(N-1)/2$ parameters (means, variances, and covariances), each defined by expectation. In fact, the total number of parameters is even smaller than this, namely $2N + N(N-1)/2$; the first term gives the number of mean and variance parameters, and the second the number of correlations.

This brings us to another useful feature of multivariate normal distributions: they are fully defined by expectations, so we need not concern ourselves with probability tables as we did when dealing with discrete variables. Conditional probabilities can be expressed as conditional expectations, and notions such as conditional independence that define the structure of graphical models can be expressed in terms of equality relationships among conditional expectations. For instance, to express the conditional independence of Y and X, given Z,

$$P(Y|X,Z) = P(Y|Z)$$

we can write

$$E[Y|X,Z] = E[Y|Z]$$

(where Z is a set of variables).

This feature of normal systems gives us an incredibly useful ability: Substituting expectations for probabilities allows us to use regression (a predictive method) to determine causal information. The next useful feature of normal distributions is their linearity: every conditional expectation $E[Y|X_1, X_2, \ldots, X_n]$ is given by a linear combination of the conditioning variables. Formally,

$$E[Y|X_1 = x_1, X_2 = x_2, \ldots, X_n = x_n] = r_0 + r_1x_1 + r_2x_2 + \cdots + r_nx_n$$

where each of the slopes $r_1, r_2, \ldots, r_n$ is a partial regression coefficient as defined in Sections 1.3.10 and 1.3.11.

The magnitudes of these slopes do not depend on the values $x_1, x_2, \ldots, x_n$ of the conditioning variables, called *regressors*; they depend only on which variables are chosen as regressors. In other words, the sensitivity of Y to the measurement $X_i = x_i$ does not depend on the measured values of the other variables in the regression; it depends only on which variables we choose to measure. It doesn't matter whether $X_i = 1, X_i = 2$, or $X_i = 312.3$; as long as we regress Y on $X_1, X_2, \ldots, X_n$ all slopes will remain the same.

This unique and useful feature of normal distributions is illustrated in Figures 1.1 and 1.2 of Chapter 1. Figure 1.1 shows that regardless of what level of age we choose, the slope of Y on X at that level is the same. If, however, we do not hold age constant (i.e., we do not regress on it), the slope becomes vastly different, as is shown in Figure 1.2.

The linearity assumption also permits us to fully specify the functions in the model by annotating the causal graph with a *path coefficient* (or structural coefficient) along each edge. The path coefficient β along the edge $X \rightarrow Y$ quantifies the contribution of X in the function that defines Y in the model. For instance, if the function defines $Y = 3X + U$, the path coefficient of $X \rightarrow Y$ will be 3. The path coefficients $\beta_1, \beta_2, \ldots, \beta_n$ are fundamentally different from the regression coefficients $r_1, r_2, \ldots, r_n$ that we discussed in Section 1.3. The former are "structural" or "causal," whereas the latter are statistical. The difference is explained in the next section.

Many of the regression methods we discuss are far more general, applying in situations where the variables $X_1, \ldots, X_k$ follow distribution far from multivariate normal; for example, when some of the X_i's are categorical or even binary. Such generalizations also therefore allow the conditional mean $E(Y|X_1 = x_1, \ldots, X_k = x_k)$ to include nonlinear combinations of the X_i's, including such terms as X_1X_2, for example, to allow for effect modification, or interaction. Since we are conditioning on the values of the X_i's, it is usually not necessary to enforce a distributional assumption for such variables. Nevertheless, the full multivariate normal scenario provides considerable insight into structural causal models.

3.8.1 Structural versus Regression Coefficients

As we are now about to deal with linear models, and thus, as a matter of course, with regression-like equations, it is of paramount importance to define the difference between regression equations and the structural equations we have used in SCMs throughout the book. A regression equation is descriptive; it makes no assumptions about causation. When we write $y = r_1x + r_2z + \epsilon$, as a regression equation, we are not saying that X and Z cause Y. We merely confess our need to know which values of r_1 and r_2 would make the equation $y = r_1x + r_2z$

the best linear approximation to the data, or, equivalently, the best linear approximation of $E(y|x, z)$.

Because of this fundamental difference between structural and regression equations, some books distinguish them by writing an arrow, instead of equality sign, in structural equations, and some distinguish the coefficients by using a different font. We distinguish them by denoting structural coefficients as α, β, and so on, and regression coefficients as r_1, r_2, and so on. In addition, we distinguish between the stochastic "error terms" that appear in these equations. Errors in regression equations are denoted ϵ_1, ϵ_2, and so on, as in Eq. (1.24), and those in structural equations by U_1, U_2, and so on, as in SCM 1.5.2. The former denote the residual errors in observation, after fitting the equation $y = r_1 x + r_2 z$ to data, whereas the latter represent latent factors (sometimes called "disturbances" or "omitted variables") that influence Y and are not themselves affected by X. The former are human-made (due to imperfect fitting); the latter are nature-made.

Though they are not causally binding themselves, regression equations are of significant use in the study of causality as it pertains to linear systems. Consider: In Section 3.2, we were able to express the effects of interventions in terms of conditional probabilities, as, for example, in the adjustment formula of Eq. (3.5). In linear systems, the role of conditional probabilities will be taken over by regression coefficients, since these coefficients represent the dependencies induced by the model and, in addition, they are easily estimable using least square analyses. Similarly, whereas the testable implications of nonparametric models are expressed in the form of conditional independencies, these independencies are signified in linear models by vanishing regression coefficients, like those discussed in Section 1.3.11. Specifically, given the regression equation

$$y = r_0 + r_1 x_1 + r_2 x_2 + \cdots + r_n x_n + \epsilon$$

if $r_i = 0$, then Y is independent of X_i conditional on all the other regression variables.

3.8.2 The Causal Interpretation of Structural Coefficients

In a linear system, every path coefficient stands for the direct effect of the independent variable, X, on the dependent variable, Y. To see why this is so, we refer to the interventional definition of direct effect given in Section 3.7 (Eq. (3.18)), which calls for computing the change in Y as X increases by one unit whereas all other parents of Y are held constant. When we apply this definition to any linear system, regardless of whether the disturbances are correlated or not, the result will be the path coefficient on the arrow $X \rightarrow Y$.

Consider, for example, the model in Figure 3.13, and assume we wish to estimate the direct effect of Z on Y. The structural equations in the fully specified model read:

$$X = U_X$$

$$Z = aX + U_Z$$

$$W = bX + cZ + U_W$$

$$Y = dZ + eW + U_Y$$

Writing Eq. (3.18) in expectation form, we obtain

$$DE = E[Y|do(Z = z + 1), do(W = w)] - E[Y|do(Z = z), do(W = w)]$$

since W is the only other parent of Y in the graph. Applying the *do* operators by deleting the appropriate equations from the model, the postincrease term in DE becomes $d(z + 1) + ew$ and the preincrease term becomes $dz + ew$. As expected, the difference between the two is d—the path coefficient between Z and Y. Note that the license to reduce the equation in this way comes directly from the definition of the *do*-operator (Eq. (3.18)) making no assumption about correlations among the U factors; the equality $DE = d$ would be valid even if the error term U_Y were correlated with U_Z, though this would have made d nonidentifiable. The same goes for the other direct effects; every structural coefficient represents a direct effect, regardless of how the error terms are distributed. Note also that variable X, as well as the coefficients a, b, and c, do not enter into this computation, because the "surgeries" required by the *do* operators remove them from the model.

That is all well and good for the direct effect. Suppose, however, we wish to calculate the *total* effect of Z on Y.

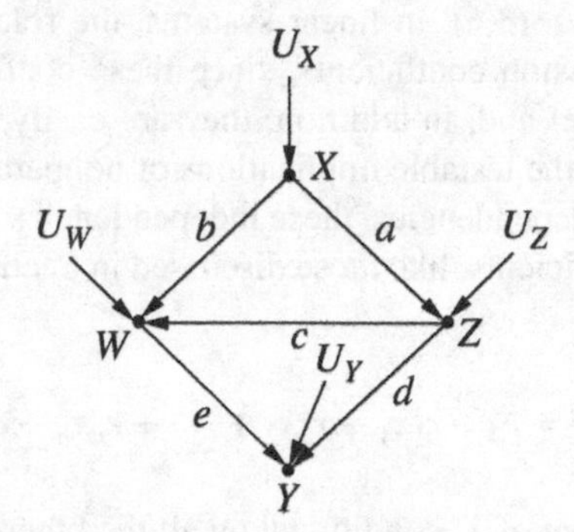

Figure 3.13 A graphical model illustrating the relationship between path coefficients and total effects

In a linear system, the total effect of X on Y is simply the sum of the products of the coefficients of the edges on every nonbackdoor path from X to Y.

That's a bit of a mouthful, so think of it as a process: To find the total effect of X on Y, first find every nonbackdoor path from X to Y; then, for each path, multiply all coefficients on the path together; then add up all the products.

The reason for this identity lies in the nature of SCMs. Consider again the graph of Figure 3.13. Since we want to find the total effect of Z on Y, we should first intervene on Z, removing all arrows going into Z, then express Y in terms of Z in the remaining model. This we can do with a little algebra:

$$\begin{aligned} Y &= dZ + eW + U_Y \\ &= dZ + e(bX + cZ) + U_Y + eU_W \\ &= (d + ec)Z + ebX + U_Y + eU_W \end{aligned}$$

The final expression is in the form $Y = \tau Z + U$, where $\tau = d + ec$ and U contains only terms that do not depend on Z in the modified model. An increase of a single unit in Z, therefore, will increase Y by τ—the definition of the total effect. A quick examination will show that τ

is the sum of the products of the coefficients on the two nonbackdoor paths from Z to Y. This will be the case in all linear models; algebra demands it. Moreover, the sum of product rule will be valid regardless of the distributions of the U variables and regardless of whether they are dependent or independent.

3.8.3 *Identifying Structural Coefficients and Causal Effect*

Thus far, we have expressed the total and direct effects in terms of path coefficients, assuming that the latter are either known to us a priori or estimated from interventional experiments. We now tackle a much harder problem; estimating total and direct effects from nonexperimental data. This problem is known as "identifiability" and, mathematically, it amounts to expressing the path coefficients associated with the total and direct effects in terms of the covariances σ_{XY} or regression coefficients $R_{YX \cdot Z}$, where X and Y are any two variables in the model, and Z a set of variables in the model (Eqs. (1.27) and (1.28) and Section 1.3.11).

In many cases, however, it turns out that to identify direct and total effects, we do not need to identify each and every structural parameter in the model. Let us first demonstrate with the total effect, τ. The backdoor criterion gives us the set Z of variables we need to adjust for in order to determine the causal effect of X on Y. How, though, do we make use of the criterion to determine effects in a linear system? In principle, once we obtain the set, Z, we can estimate the conditional expectation of Y given X and Z and, then, averaging over Z, we can use the resultant dependence between Y and X to measure the effect of X on Y. We need only translate this procedure to the language of regression.

The translation is rather simple. First, we find a set of covariates Z that satisfies the backdoor criterion from X to Y in the model. Then, we regress Y on X and Z. The coefficient of X in the resulting equation represents the true causal effect of X on Y. The reasoning for this is similar to the reasoning we used to justify the backdoor criterion in the first place—regressing on Z adds those variables into the equation, blocking all backdoor paths from X and Y, thus preventing the coefficient of X from absorbing the spurious information those paths contain.

For example, consider a linear model that complies with the graph in Figure 3.14. If we want to find the total causal effect of X on Y, we first determine, using the backdoor criterion, that we must adjust for T. So we regress Y on X and T, using the regression equation $y = r_X X +$

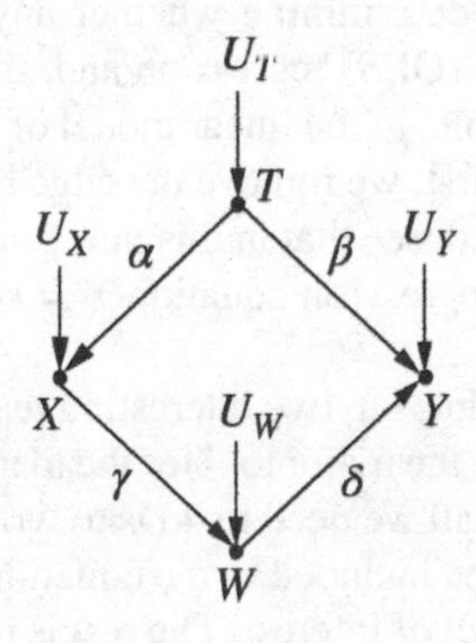

Figure 3.14 A graphical model in which X has no direct effect on Y, but a total effect that is determined by adjusting for T

$r_T T + \epsilon$. The coefficient r_X represents the total effect of X on Y. Note that this identification was possible without identifying any of the model parameters and without measuring variable W; the graph structure in itself gave us the license to ignore W, regress Y on T and X only, and identify the total effect (of X on Y) with the coefficient of X in that regression.

Suppose now that instead of the total causal effect, we want to find X's direct effect on Y. In a linear system, this direct effect is the structural coefficient α in the function $y = \alpha x + \beta z + \cdots + U_Y$ that defines Y in the system. We know from the graph of Figure 3.14 that $\alpha = 0$, because there is no direct arrow from X to Y. So, in this particular case, the answer is trivial: the direct effect is zero. But in general, how do we find the magnitude of α from data, if the model does not determine its value?

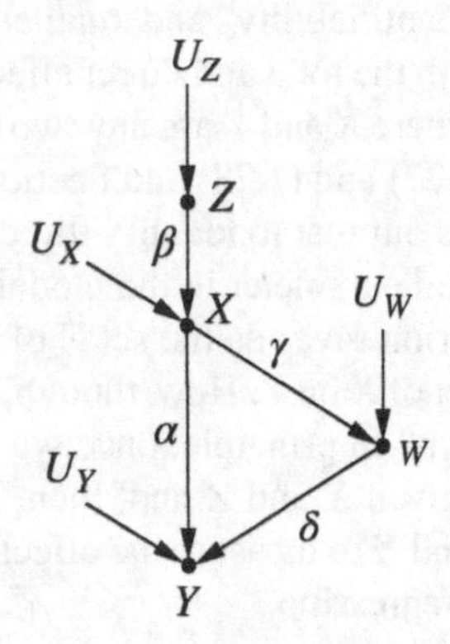

Figure 3.15 A graphical model in which X has direct effect α on Y

We can invoke a procedure similar to backdoor, except that now, we need to block not only backdoor paths but also indirect paths going from X to Y. First, we remove the edge from X to Y (if such an edge exists), and call the resulting graph G_α. If, in G_α, there is a set of variables Z that d-separates X and Y, then we can simply regress Y on X and Z. The coefficient of X in the resulting equation will equal the structural coefficient α.

The procedure above, which we might as well call "The Regression Rule for Identification" provides us with a quick way of determining whether any given parameter (say α) can be identified by ordinary least square (OLS) regression and, if so, what variables should go into the regression equation. For example, in the linear model of Figure 3.15, we can find the direct effect of X on Y by this method. First, we remove the edge between X and Y and get the graph G_α shown in Figure 3.16. It's easy to see that in this new graph, W d-separates X and Y. So we regress Y on X and W, using the regression equation $Y = r_X X + r_W W + \epsilon$. The coefficient r_X is the direct effect of X on Y.

Summarizing our observations thus far, two interesting features emerge. First, we see that, in linear systems, regression serves as the major tool for the identification and estimation of causal effects. To estimate a given effect, all we need to do is to write down a regression equation and specify (1) what variables should be included in the equation and (2) which of the coefficients in that equation represents the effect of interest. The rest is routine least square analysis on the sampled data which, as we remarked before, is facilitated by a variety of extremely efficient software packages. Second, we see that, as long as the U variables are independent of each

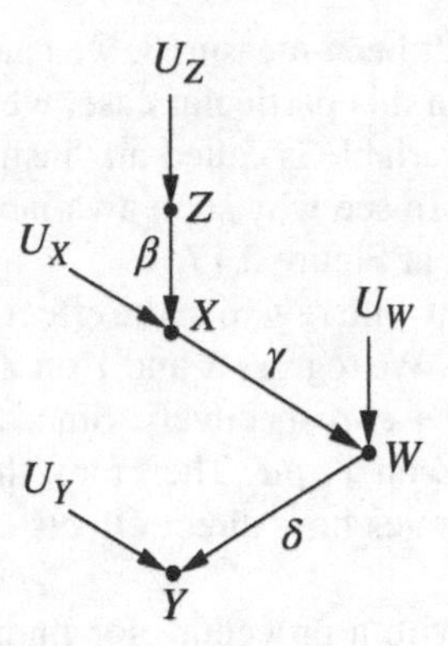

Figure 3.16 By removing the direct edge from X to Y and finding the set of variables $\{W\}$ that d-separate them, we find the variables we need to adjust for to determine the direct effect of X on Y

other, and all variables in the graph are measured, every structural parameter can be identified in this manner, namely, there is at least one identifying regression equation in which one of the coefficients corresponds to the parameter we seek to estimate. One such equation is obviously the structural equation itself, with the parents of Y serving as regressors. But there may be several other identifying equations, with possibly better features for estimation, graphical analysis can reveal them all (see Study question 3.8.1(c)). Moreover, when some variables are not measured, or when some error terms are correlated, the task of finding an identifying regression from the structural equations themselves would normally be insurmountable; the G_α procedure then becomes indispensable (see Study question 3.8.1(d)).

Remarkably, the regression rule procedure has eluded investigators for almost a century, possibly because it is extremely difficult to articulate in algebraic, nongraphical terms.

Suppose, however, there is no set of variables that d-separates X and Y in G_α. For instance, in Figure 3.17, X and Y have an unobserved common cause represented by the dashed

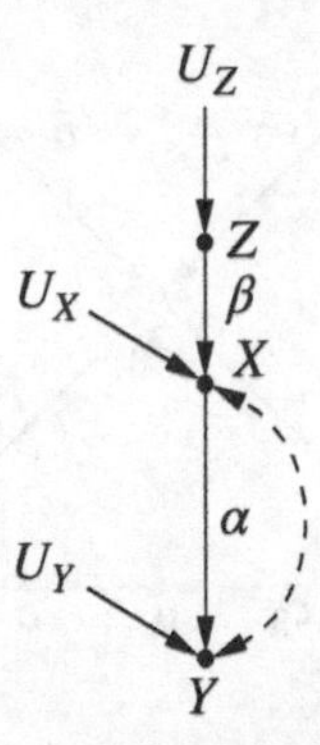

Figure 3.17 A graphical model in which we cannot find the direct effect of X on Y via adjustment, because the dashed double-arrow arc represents the presence of a backdoor path between X and Y, consisting of unmeasured variables. In this case, Z is an instrument with regard to the effect of X on Y that enables the identification of α

double-arrowed arc. Since it hasn't been measured, we can't condition on it, so X and Y will always be dependent through it. In this particular case, we may use an *instrumental variable* to determine the direct effect. A variable is called an "instrument" if it is d-separated from Y in G_α and it is d-connected to X. To see why such a variable enables us to identify structural coefficients, we take a closer look at Figure 3.17.

In Figure 3.17, Z is an instrument with regard to the effect of X on Y because it is d-connected to X and d-separated from Y in G_α. We regress X and Y on Z separately, yielding the regression equations $y = r_1 z + \epsilon$ and $x = r_2 z + \epsilon$, respectively. Since Z emits no backdoors, r_2 equals β and r_1 equals the total effect of Z on Y, $\beta\alpha$. Therefore, the ratio r_1/r_2 provides the desired coefficient α. This example illustrates how direct effects can be identified from total effects but not the other way around.

Graphical models provide us with a procedure for finding all instrumental variables in a system, though the procedure for enumerating them is beyond the scope of this book. Those interested in learning more can (see Chen and Pearl 2014; Kyono 2010).

Study questions

Study question 3.8.1

Model 3.1

$$Y = aW_3 + bZ_3 + cW_2 + U \qquad X = t_1 W_1 + t_2 Z_3 + U'$$
$$W_3 = c_3 X + U_3' \qquad W_1 = a_1' Z_1 + U_1'$$
$$Z_3 = a_3 Z_1 + b_3 Z_2 + U_3 \qquad Z_1 = U_1$$
$$W_2 = c_2 Z_2 + U_2' \qquad Z_2 = U_2$$

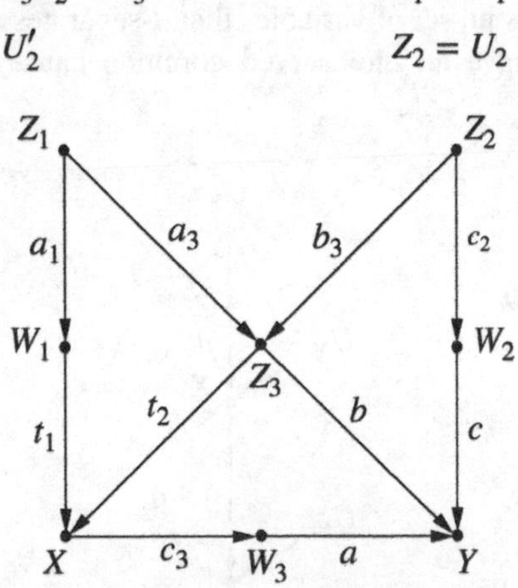

Figure 3.18 Graph corresponding to Model 3.1 in Study question 3.8.1

Given the model depicted above, answer the following questions:

(All answers should be given in terms of regression coefficients in specified regression equations.)

(a) Identify three testable implications of this model.
(b) Identify a testable implication assuming that only X, Y, W_3, and Z_3 are observed.
(c) For each of the parameters in the model, write a regression equation in which one of the coefficients is equal to that parameter. Identify the parameters for which more than one such equation exists.
(d) Suppose X, Y, and W_3 are the only variables observed. Which parameters can be identified from the data? Can the total effect of X on Y be estimated?
(e) If we regress Z_1 on all other variables in the model, which regression coefficient will be zero?
(f) The model in Figure 3.18 implies that certain regression coefficients will remain invariant when an additional variable is added as a regressor. Identify five such coefficients with their added regressors.
(g) Assume that variables Z_2 and W_2 cannot be measured. Find a way to estimate b using regression coefficients. [Hint: Find a way to turn Z_1 into an instrumental variable for b.]

3.8.4 Mediation in Linear Systems

When we can assume linear relationships between variables, mediation analysis becomes much simpler than the analysis conducted in nonlinear or nonparametric systems (Section 3.7). Estimating the direct effect of X on Y, for instance, amounts to estimating the path coefficient between the two variables, and this reduces to estimating correlation coefficients, using the techniques introduced in Section 3.8.3. The indirect effect, similarly, is computed via the difference $IE = \tau - DE$, where τ, the total effect, can be estimated by regression in the manner shown in Figure 3.14. In nonlinear systems, on the other hand, the direct effect is defined through expressions such as (3.18), or

$$DE = E[Y|do(x,z)] - E[Y|do(x',z)]$$

where $Z = z$ represents a specific stratum of all other parents of Y (besides X). Even when the identification conditions are satisfied, and we are able to reduce the *do*() operators (by adjustments) to ordinary conditional expectations, the result will still depend on the specific values of x, x', and z. Moreover, the indirect effect cannot be given a definition in terms as *do*-expressions, since we cannot disable the capacity of Y to respond to X by holding variables constant. Nor can the indirect effect be defined as the difference between the total and direct effects, since differences do not faithfully reflect operations in nonlinear systems to X.

Such an operation will be introduced in Chapter 4 (Sections 4.4.5 and 4.5.2) using the language of counterfactuals.

Bibliographical Notes for Chapter 3

Study question 3.3.2 is a version of Lord's paradox (Lord 1967), and is described in Glymour (2006), Hernández-Díaz et al. (2006), Senn (2006), and Wainer (1991). A unifying treatment is given in Pearl (2016). The definition of the *do*-operator and "ACE" in terms of a modified model, has its conceptual origin with the economist Trygve Haavelmo (1943), who was the first to simulate interventions by modifying equations in the model (see Pearl (2015c) for historical account). Strotz and Wold (1960) later advocated "wiping out" the equation determining X, and Spirtes et al. (1993) gave it a graphical representation in a form of a "manipulated graph." The "adjustment formula" of Eq. (3.5) as well as the "truncated product formula" first appeared in Spirtes et al. (1993), though these are implicit in the G-computation formula of Robins (1986), which was derived using counterfactual assumptions (see Chapter 4). The backdoor criterion of Definition 3.3.1 and its implications for adjustments were introduced in Pearl (1993). The front-door criterion and a general calculus for identifying causal effects (named *do*-calculus) from observations and experimental data were introduced in Pearl (1995) and were further improved in Tian and Pearl (2002), Shpitser and Pearl (2007), and Bareinboim and Pearl (2012). Section 3.5, and the identification of conditional interventions and c-specific effects is based on (Pearl 2009, pp. 113–114). Its extension to dynamic, time-varying policies is described in Pearl and Robins (1995) and (Pearl 2009, pp. 119–126). More recently, the *do*-calculus was used to solve problems of external validity, data-fusion, and meta-analysis (Bareinboim and Pearl 2013, Bareinboim and Pearl 2016, and Pearl and Bareinboim 2014). The role of covariate-specific effects in assessing interaction, moderation or effect modification is described in Morgan and Winship (2014) and Vanderweele (2015), whereas applications of Rule 2 to the detection of latent heterogeneity are described in Pearl (2015b). Additional discussions on the use of inverse probability weighting (Section 3.6) can be found in Hernán and Robins (2006). Our discussion of mediation (Section 3.7) and the identification of CDEs are based on Pearl (2009, pp. 126–130), whereas the fallibility of "conditioning" on a mediator to assess direct effects is demonstrated in Pearl (1998) as well as Cole and Hernán (2002).

The analysis of mediation has become extremely active in the past 15 years, primarily due to the advent of counterfactual logic (see Section 4.4.5); a comprehensive account of this progress is given in Vanderweele (2015). A tutorial survey of causal inference in linear systems (Section 3.8), focusing on parameter identification, is provided by Chen and Pearl (2014). Additional discussion on the confusion of regression versus structural equations can be found in Bollen and Pearl (2013).

A classic, and still the best textbook on the relationships between structural and regession coefficients is Heise (1975) (available online: http://www.indiana.edu/~socpsy/public_files/CausalAnalysis.zip). Other classics are Duncan (1975), Kenny (1979), and Bollen (1989). Classical texts, however, fall short of providing graphical tools of identification, such as those invoking backdoor and G_α (see Study question 3.8.1). A recent exception is Kline (2016).

Introductions to instrumental variables can be found in Greenland (2000) and in many textbooks of econometrics (e.g., Bowden and Turkington 1984, Wooldridge 2013). *The Book of Why* (Pearl and Mackenzie 2018) traces the idea to John Snow (1813–1858) and his investigation of the Cholera epidemic of 1853–54. Generalized instrumental variables, extending the classical definition of Section 3.8.3 were introduced in Brito and Pearl (2002).

The program DAGitty (which is available online: http://www.dagitty.net/dags.html), permits users to search the graph for generalized instrumental variables, and reports the resulting IV estimators (Textor et al. 2011).

4

Counterfactuals and Their Applications

4.1 Counterfactuals

While driving home last night, I came to a fork in the road, where I had to make a choice: to take the freeway ($X = 1$) or go on a surface street named Sepulveda Boulevard ($X = 0$). I took Sepulveda, only to find out that the traffic was touch and go. As I arrived home, an hour later, I said to myself: "Gee, I should have taken the freeway."

What does it mean to say, "I should have taken the freeway"? Colloquially, it means, "If I had taken the freeway, I would have gotten home earlier." Scientifically, it means that my mental estimate of the expected driving time on the freeway, on that same day, under the identical circumstances, and governed by the same idiosyncratic driving habits that I have, would have been lower than my actual driving time.

This kind of statement—an "if" statement in which the "if" portion is untrue or unrealized—is known as a *counterfactual*. The "if" portion of a counterfactual is called the *hypothetical condition*, or more often, the *antecedent*. We use counterfactuals to emphasize our wish to compare two outcomes (e.g., driving times) under the exact same conditions, differing only in one aspect: the antecedent, which in our case stands for "taking the freeway" as opposed to the surface street. The fact that we know the outcome of our actual decision is important, because my estimated driving time on the freeway after seeing the consequences of my actual decision (to take Sepulveda) may be totally different from my estimate prior to seeing the consequence. The consequence (1 hour) may provide valuable evidence for the assessment, for example, that the traffic was particularly heavy on that day, and that it might have been due to a brush fire. My statement "I should have taken the freeway" conveys the judgment that whatever mechanisms impeded my speed on Sepulveda would not have affected the speed on the freeway to the same extent. My retrospective estimate is that a freeway drive would have taken less than 1 hour, and this estimate is clearly different than my prospective estimate was, when I made the decision prior to seeing the consequences—otherwise, I would have taken the freeway to begin with.

Causal Inference in Statistics: A Primer, First Edition. Judea Pearl, Madelyn Glymour, and Nicholas P. Jewell.

Companion Website: www.wiley.com/go/Pearl/Causality

If we try to express this estimate using *do*-expressions, we come to an impasse. Writing

$$E(\textit{driving time}|\textit{do}(\textit{freeway}), \textit{driving time} = 1 \textit{ hour})$$

leads to a clash between the driving time we wish to estimate and the actual driving time observed. Clearly, to avoid this clash, we must distinguish symbolically between the following two variables:

1. Actual driving time
2. Hypothetical driving time under freeway conditions when actual surface driving time is known to be 1 hour.

Unfortunately, the *do*-operator is too crude to make this distinction. While the *do*-operator allows us to distinguish between two probabilities, $P(\textit{driving time} = T \mid \textit{do}(\textit{freeway}))$ and $P(\textit{driving time} = T \mid \textit{do}(\textit{Sepulveda}))$, it does not offer us the means of distinguishing between the two variables themselves, one standing for the time on Sepulveda, the other for the hypothetical time on the freeway. We need this distinction in order to let the actual driving time (on Sepulveda) inform our assessment of the hypothetical driving time.

Fortunately, making this distinction is easy; we simply use different subscripts to label the two outcomes. We denote the freeway driving time by $Y_{X=1}$ (or Y_1, where context permits) and Sepulveda driving time by $Y_{X=0}$ (or Y_0). In our case, since Y_0 is the Y actually observed, the quantity we wish to estimate is

$$E(Y_{X=1}|X = 0, Y = Y_0 = 1) \tag{4.1}$$

The novice student may feel somewhat uncomfortable at the sight of the last expression, which contains an eclectic mixture of three variables: one hypothetical and two observed, with the hypothetical variable $Y_{X=1}$ predicated upon one event ($X = 1$) and conditioned upon the conflicting event, $X = 0$, which was actually observed. We have not encountered such a clash before. When we used the *do*-operator to predict the effect of interventions, we wrote expressions such as

$$E[Y|do(X = x)] \tag{4.2}$$

The Y in this expression is predicated upon the event $X = x$. With our new notation, the expression might as well have been written $E[Y_{X=x}]$. But since all variables in this expression were measured in the same world, there is no need to abandon the *do*-operator and invoke counterfactual notation.

We run into problems with counterfactual expressions like (4.1) because $Y_{X=1} = y$ and $X = 0$ are—and must be—events occurring under different conditions, sometimes referred to as "different worlds." This problem does not occur in intervention expressions, because Eq. (4.1) seeks to estimate our total drive time in a world where we chose the freeway, given that the actual drive time (in the world where we chose Sepulveda) was 1 hour, whereas Eq. (4.2) seeks to estimate the expected drive time in a world where we chose the freeway, with no reference whatsoever to another world.

In Eq. (4.1), however, the clash prevents us from reducing the expression to a *do*-expression, which means that it cannot be estimated from interventional experiments. Indeed, a randomized controlled experiment on the two decision options will never get us the estimate we want. Such experiments can give us $E[Y_1] = E[Y|do(freeway)]$ and $E[Y_0] = E[Y|do(Sepulveda)]$, but the fact that we cannot take both the freeway and Sepulveda simultaneously prohibits us from estimating the quantity we wish to estimate, that is, the conditional expectation $E[Y_1|X = 0, Y = 1]$. One might be tempted to circumvent this difficulty by measuring the freeway time at a later time, or of another driver, but then conditions may change with time, and the other driver may have different driving habits than I. In either case, the driving time we would be measuring under such surrogates will only be an approximation of the one we set out to estimate, Y_1, and the degree of approximation would vary with the assumptions we can make on how similar those surrogate conditions are to my own driving time had I taken the freeway. Such approximations may be appropriate for estimating the target quantity under some circumstances, but they are not appropriate for *defining* it. Definitions should accurately capture what we wish to estimate, and for this reason, we must resort to a subscript notation, Y_1, with the understanding that Y_1 is my "would-be" driving time, had I chosen the freeway at that very juncture of history.

Readers will be pleased to know that their discomfort with the clashing nature of Eq. (4.1) will be short-lived. Despite the hypothetical nature of the counterfactual Y_1, the structural causal models that we have studied in Part Two of the book will prove capable not only of computing probabilities of counterfactuals for any fully specified model, but also of estimating those probabilities from data, when the underlying functions are not specified or when some of the variables are unmeasured.

In the next section, we detail the methods for computing and estimating properties of counterfactuals. Once we have done that, we'll use those methods to solve all sorts of complex, seemingly intractable problems. We'll use counterfactuals to determine the efficacy of a job training program by figuring out how many enrollees would have gotten jobs had they not enrolled; to predict the effect of an additive intervention (adding 5 mg/l of insulin to a group of patients with varying insulin levels) from experimental studies that exercised a uniform intervention (setting the group of patients' insulin levels to the same constant value); to ascertain the likelihood that an individual cancer patient would have had a different outcome, had she chosen a different treatment; to prove, with a sufficient probability, whether a company was discriminating when they passed over a job applicant; and to suss out, via analysis of direct and indirect effects, the efficacy of gender-blind hiring practices on rectifying gender disparities in the workforce.

All this and more, we can do with counterfactuals. But first, we have to learn how to define them, how to compute them, and how to use them in practice.

4.2 Defining and Computing Counterfactuals

4.2.1 The Structural Interpretation of Counterfactuals

We saw in the subsection on interventions that structural causal models can be used to predict the effect of actions and policies that have never been implemented before. The action of setting a variable, X, to value x is simulated by replacing the structural equation for X with the equation $X = x$. In this section, we show that by using the same operation in a slightly different context, we can use SCMs to define what counterfactuals stand for, how to read counterfactuals from a given model, and how probabilities of counterfactuals can be estimated when portions of the models are unknown.

We begin with a fully specified model M, for which we know both the functions $\{F\}$ and the values of all exogenous variables. In such a deterministic model, every assignment $U = u$ to the exogenous variables corresponds to a single member of, or "unit" in a population, or to a "situation" in nature. The reason for this correspondence is as follows: Each assignment $U = u$ uniquely determines the values of all variables in V. Analogously, the characteristics of each individual "unit" in a population have unique values, depending on that individual's identity. If the population is "people," these characteristics include salary, address, education, propensity to engage in musical activity, and all other properties we associate with that individual at any given time. If the population is "agricultural lots," these characteristics include soil content, surrounding climate, and local wildlife, among others. There are so many of these defining properties that they cannot all possibly be included in the model, but taken all together, they uniquely distinguish each individual and determine the values of the variables we do include in the model. It is in this sense that every assignment $U = u$ corresponds to a single member or "unit" in a population, or to a "situation" in nature.

For example, if $U = u$ stands for the defining characteristics of an individual named Joe, and X stands for a variable named "salary," then $X(u)$ stands for Joe's salary. If $U = u$ stands for the identity of an agricultural lot and Y stands for the yield measured in a given season, then $Y(u)$, stands for the yield produced by lot $U = u$ in that season.

Consider now the counterfactual sentence, "Y would be y had X been x, in situation $U = u$," denoted $Y_x(u) = y$, where Y and X are any two variables in V. The key to interpreting such a sentence is to treat the phrase "had X been x" as an instruction to make a minimal modification in the current model so as to establish the antecedent condition $X = x$, which is likely to conflict with the observed value of $X, X(u)$. Such a minimal modification amounts to replacing the equation for X with a constant x, which may be thought of as an external intervention $do(X = x)$, not necessarily by a human experimenter. This replacement permits the constant x to differ from the actual value of X (namely, $X(u)$) without rendering the system of equations inconsistent, and in this way, it allows all variables, exogenous as well as endogenous, to serve as antecedents to other variables.

We demonstrate this definition on a simple causal model consisting of just three variables, X, Y, U, and defined by two equations:

$$X = aU \tag{4.3}$$

$$Y = bX + U \tag{4.4}$$

We first compute the counterfactual $Y_x(u)$, that is, what Y would be had X been x, in situation $U = u$. Replacing the first equation with $X = x$ gives the "modified" model M_x:

$$X = x$$

$$Y = bX + U$$

Substituting $U = u$ and solving for Y gives

$$Y_x(u) = bx + u$$

Table 4.1 The values attained by $X(u)$, $Y(u)$, $Y_x(u)$, and $X_y(u)$ in the linear model of Eqs. (4.3) and (4.4)

u	$X(u)$	$Y(u)$	$Y_1(u)$	$Y_2(u)$	$Y_3(u)$	$X_1(u)$	$X_2(u)$	$X_3(u)$
1	1	2	2	3	4	1	1	1
2	2	4	3	4	5	2	2	2
3	3	6	4	5	6	3	3	3

which is expected, since the meaning of the structural equation $Y = bX + U$ is, exactly "the value that Nature assigns to Y must be U plus b times the value assigned to X." To demonstrate a less obvious result, let us examine the counterfactual $X_y(u)$, that is, what X would be had Y been y in situation $U = u$. Here, we replace the second equation by the constant $Y = y$ and, solving for X, we get $X_y(u) = au$, which means that X remains unaltered by the hypothetical condition "had Y been y." This should be expected, if we interpret this hypothetical condition as emanating from an external, albeit unspecified, intervention. It is less expected if we do not invoke the intervention metaphor but merely treat $Y = y$ as a spontaneous, unanticipated change. The invariance of X under such a counterfactual condition reflects the intuition that hypothesizing future eventualities does not alter the past.

Each SCM encodes within it many such counterfactuals, corresponding to the various values that its variables can take. To illustrate additional counterfactuals generated by this model, let us assume that U can take on three values, 1, 2, and 3, and let $a = b = 1$ in Eqs. (4.3) and (4.4). Table 4.1 gives the values of $X(u)$, $Y(u)$, $Y_x(u)$, and $X_y(u)$ for several levels of x and y. For example, to compute $Y_2(u)$ for $u = 2$, we simply solve a new set of equations, with $X = 2$ replacing $X = aU$, and obtain $Y_2(u) = 2 + u = 4$. The computation is extremely simple, which goes to show that, while counterfactuals are considered hypothetical, or even mystical from a statistical view point, they emerge quite naturally from our perception of reality, as encoded in structural models. Every structural equation model assigns a definitive value to every conceivable counterfactual.

From this example, the reader may get the impression that counterfactuals are no different than ordinary interventions, captured by the *do*-operator. Note, however, that, in this example we computed not merely the probability or expected value of Y under one intervention or another, but the actual value of Y under the hypothesized new condition $X = x$. For each situation $U = u$, we obtained a definite number, $Y_x(u)$, which stands for that hypothetical value of Y in that situation. The *do*-operator, on the other hand, is only defined on probability distributions and, after deleting the factor $P(x_i|pa_i)$ from the product decomposition (Eq. (1.29)), always delivers probabilistic results such as $E[Y|do(x)]$. From an experimentalist perspective, this difference reflects a profound gap between population and individual levels of analysis; the *do*(x)-operator captures the behavior of a population under intervention, whereas $Y_x(u)$ describes the behavior of a specific individual, $U = u$, under such interventions. This difference has far-reaching consequences, and will enable us to define probabilities of concepts such as credit, blame, and regret, which the *do*-operator is not able to capture.

4.2.2 *The Fundamental Law of Counterfactuals*

We are now ready to generalize the concept of counterfactuals to any structural model, M. Consider any arbitrary two variables X and Y, not necessarily connected by a single equation.

Let M_x stand for the modified version of M, with the equation of X replaced by $X = x$. The formal definition of the counterfactual $Y_x(u)$ reads

$$Y_x(u) = Y_{M_x}(u) \tag{4.5}$$

In words: The counterfactual $Y_x(u)$ in model M is defined as the solution for Y in the "surgically modified" submodel M_x. Equation (4.5) is one of the most fundamental principles of causal inference. It allows us to take our scientific conception of reality, M, and use it to generate answers to an enormous number of hypothetical questions of the type "What would Y be had X been x?" The same definition is applicable when X and Y are sets of variables, if by M_x we mean a model where the equations of all members of X are replaced by constants. This raises enormously the number of counterfactual sentences computable by a given model and brings up an interesting question: How can a simple model, consisting of just a few equations, assign values to so many counterfactuals? The answer is that the values that these counterfactuals receive are not totally arbitrary, but must cohere with each other to be consistent with an underlying model.

For example, if we observe $X(u) = 1$ and $Y(u) = 0$, then $Y_{X=1}(u)$ must be zero, because setting X to a value it already has, $X(u)$, should produce no change in the world. Hence, Y should stay at its current value of $Y(u) = 0$.

In general, counterfactuals obey the following *consistency rule*:

$$\textit{if} \quad X = x \quad \textit{then} \quad Y_x = Y \tag{4.6}$$

If X is binary, then the consistency rule takes the convenient form:

$$Y = XY_1 + (1 - X)Y_0$$

which can be interpreted as follows: Y_1 is equal to the observed value of Y whenever X takes the value one. Symmetrically, Y_0 is equal to the observed value of Y whenever X is zero. All these constraints are automatically satisfied if we compute counterfactuals through Eq. (4.5).

4.2.3 *From Population Data to Individual Behavior—An Illustration*

To illustrate the use of counterfactuals in reasoning about the behavior of an individual unit, we refer to the model depicted in Figure 4.1, which represents an "encouragement design": X represents the amount of time a student spends in an after-school remedial program, H the amount of homework a student does, and Y a student's score on the exam. The value of each variable is given as the number of standard deviations above the mean the student falls (i.e.,

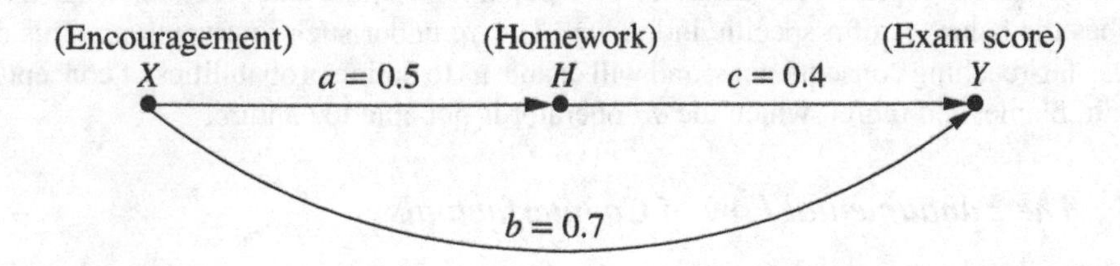

Figure 4.1 A model depicting the effect of Encouragement (X) on student's score

the model is *standardized* so that all variables have mean 0 and variance 1). For example, if $Y = 1$, then the student scored 1 standard deviation above the mean on his or her exam. This model represents a randomized pilot program, in which students are assigned to the remedial sessions by the luck of the draw.

Model 4.1

$$X = U_X$$
$$H = a \cdot X + U_H$$
$$Y = b \cdot X + c \cdot H + U_Y$$
$$\sigma_{U_i U_j} = 0 \quad \text{for all } i, j \in \{X, H, Y\}$$

We assume that all U factors are independent and that we are given the values for the coefficients of Model 4.1 (these can be estimated from population data):

$$a = 0.5, \quad b = 0.7, \quad c = 0.4$$

Let us consider a student named Joe, for whom we measure $X = 0.5, H = 1$, and $Y = 1.5$. Suppose we wish to answer the following query: What would Joe's score have been had he doubled his study time?

In a linear SCM, the value of each variable is determined by the coefficients and the U variables; the latter account for all variation among individuals. As a result, we can use the evidence $X = 0.5, H = 1$, and $Y = 1.5$ to determine the values of the U variables associated with Joe. These values are invariant to hypothetical actions (or "miracles") such as those that might cause Joe to double his homework.

In this case, we are able to obtain the specific characteristics of Joe from the evidence:

$$U_X = 0.5,$$
$$U_H = 1 - 0.5 \cdot 0.5 = 0.75, \text{ and}$$
$$U_Y = 1.5 - 0.7 \cdot 0.5 - 0.4 \cdot 1 = 0.75.$$

Next, we simulate the action of doubling Joe's study time by replacing the structural equation for H with the constant $H = 2$. The modified model is depicted in Figure 4.2. Finally, we compute the value of Y in our modified model using the updated U values, giving

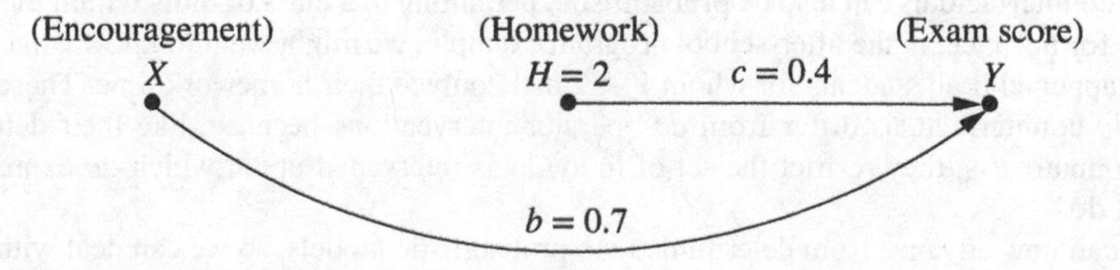

Figure 4.2 Answering a counterfactual question about a specific student's score, predicated on the assumption that homework would have increased to $H = 2$

$$\begin{aligned} Y_{H=2}(U_X = 0.5, U_H = 0.75, U_Y = 0.75) \\ &= 0.5 \cdot 0.7 + 2.0 \cdot 0.4 + 0.75 \\ &= 1.90 \end{aligned}$$

We thus conclude that Joe's score, had he doubled his homework, would have been 1.9 instead of 1.5. This, according to our convention, would mean an increase to 1.9 standard deviations above the mean, instead of the current 1.5.

In summary, we first applied the evidence $X = 0.5, H = 1$, and $Y = 1.5$ to update the values for the U variables. We then simulated an external intervention to force the condition $H = 2$ by replacing the structural equation $H = aX + U_H$ with the equation $H = 2$. Finally, we computed the value of Y given the structural equations and the updated U values. (In all of the above, we, of course, assumed that the U variables are unchanged by the hypothetical intervention on H.)

4.2.4 The Three Steps in Computing Counterfactuals

The case of Joe and the after-school program illustrates the way in which the fundamental definition of counterfactuals can be turned into a process for obtaining the value of a given counterfactual. There is a three-step process for computing any deterministic counterfactual:

(i) Abduction: Use evidence $E = e$ to determine the value of U.
(ii) Action: Modify the model, M, by removing the structural equations for the variables in X and replacing them with the appropriate functions $X = x$, to obtain the modified model, M_x.
(iii) Prediction: Use the modified model, M_x, and the value of U to compute the value of Y, the consequence of the counterfactual.

In temporal metaphors, Step (i) explains the past (U) in light of the current evidence e; Step (ii) bends the course of history (minimally) to comply with the hypothetical antecedent $X = x$; finally, Step (iii) predicts the future (Y) based on our new understanding of the past and our newly established condition, $X = x$.

This process will solve any deterministic counterfactual, that is, counterfactuals pertaining to a single unit of the population in which we know the value of every relevant variable. Structural equation models are able to answer counterfactual queries of this nature because each equation represents the mechanism by which a variable obtains its values. If we know these mechanisms, we should also be able to predict what values would be obtained had some of these mechanisms been altered, given the alterations. As a result, it is natural to view counterfactuals as *derived* properties of structural equations. (In some frameworks, counterfactuals are taken as primitives (Holland 1986; Rubin 1974).)

But counterfactuals can also be probabilistic, pertaining to a class of units within the population; for instance, in the after-school program example, we might want to know what would have happened if all students for whom $Y < 2$ had doubled their homework time. These probabilistic counterfactuals differ from *do*-operator interventions because, like their deterministic counterparts, they restrict the set of individuals intervened upon, which *do*-expressions cannot do.

We can now advance from deterministic to probabilistic models, so we can deal with questions about probabilities and expectations of counterfactuals. For example, suppose Joe is a student participating in the study of Figure 4.1, who scored $Y = y$ in the exam. What is the probability that Joe's score would be $Y = y'$ had he had five more hours of encouragement training? Or, what would his *expected* score be in such hypothetical world? Unlike in the example of Model 4.1, we now do not have information on all three variables, $\{X, Y, H\}$, and we cannot therefore determine uniquely the value u that pertains to Joe. Instead, Joe may belong to a large class of units compatible with the evidence available, each having a different value of u.

Nondeterminism enters causal models by assigning probabilities $P(U = u)$ over the exogenous variables U. These represent our uncertainty as to the identity of the subject under consideration or, when the subject is known, what other characteristics that subject has that might have bearing on our problem.

The exogenous probability $P(U = u)$ induces a unique probability distribution on the endogenous variables $V, P(v)$, with the help of which we can define and compute not only the probability of any single counterfactual, $Y_x = y$, but also the joint distributions of all combinations of observed and counterfactual variables. For example, we can determine $P(Y_x = y, Z_w = z, X = x')$, where X, Y, Z, and W are arbitrary variables in a model. Such joint probabilities refer to the proportion of individuals u in the population for which all the events in the parentheses are true, namely, $Y_x(u) = y$ and $Z_w(u) = z$ and $X(u) = x'$, allowing, in particular, w or x' to conflict with x.

A typical query about these probabilities asks, "Given that we observe feature $E = e$ for a given individual, what would we expect the value of Y for that individual to be if X had been x?" This expectation is denoted $E[Y_{X=x}|E = e]$, where we allow $E = e$ to conflict with the antecedent $X = x$. $E = e$ after the conditioning bar represents all information (or evidence) we might have about the individual, potentially including the values of X, Y, or any other variable, as we have seen in Eq. (4.1). The subscript $X = x$ represents the antecedent specified by the counterfactual sentence.

The specifics of how these probabilities and expectations are dealt with will be examined in the following sections, but for now, it is important to know that using them, we can generalize our three-step process to any probabilistic nonlinear system.

Given an arbitrary counterfactual of the form, $E[Y_{X=x}|E = e]$, the three-step process reads:

(i) **Abduction**: Update $P(U)$ by the evidence to obtain $P(U|E = e)$.
(ii) **Action**: Modify the model, M, by removing the structural equations for the variables in X and replacing them with the appropriate functions $X = x$, to obtain the modified model, M_x.
(iii) **Prediction**: Use the modified model, M_x, and the updated probabilities over the U variables, $P(U|E = e)$, to compute the expectation of Y, the consequence of the counterfactual.

We shall see in Section 4.4 that the above probabilistic procedure applies not only to retrospective counterfactual queries (queries of the form "What would have been the value of Y had X been x?") but also to certain kinds of intervention queries. In particular, it applies when we make every individual take an action that depends on the current value of his/her X. A typical example would be "additive intervention": for example, adding 5 mg/l of insulin to every patient's regiment, regardless of their previous dosage. Since the final level of insulin varies from patient to patient, this policy cannot be represented in *do*-notation.

For another example, suppose we wish to estimate, using Figure 4.1, the effect on test score provided by a school policy that sends students who are lazy on their homework ($H \leq H_0$) to attend the after-school program for $X = 1$. We can't simply intervene on X to set it equal to 1 in cases where H is low, because in our model, X is one of the causes of H.

Instead, we express the expected value of this quantity in counterfactual notation as $E[Y_{X=1}|H \leq H_0]$, which can, in principle, be computed using the above three-step method. Counterfactual reasoning and the above procedure are necessary for estimating the effect of actions and policies on subsets of the population characterized by features that, in themselves, are affected by the policy (e.g., $H \leq H_0$).

4.3 Nondeterministic Counterfactuals

4.3.1 Probabilities of Counterfactuals

To examine how nondeterminism is reflected in the calculation of counterfactuals, let us assign probabilities to the values of U in the model of Eqs. (4.3) and (4.4). Imagine that $U = \{1, 2, 3\}$ represents three types of individuals in a population, occurring with probabilities

$$P(U = 1) = \frac{1}{2}, P(U = 2) = \frac{1}{3}, \quad \text{and} \quad P(U = 3) = \frac{1}{6}$$

All individuals within a population type have the same values of the counterfactuals, as specified by the corresponding rows in Table 4.1. With these values, we can compute the probability that the counterfactuals will satisfy a specified condition. For instance, we can compute the proportion of units for which Y would be 3 had X been 2, or $Y_2(u) = 3$. This condition occurs only in the first row of the table and, since it is a property of $U = 1$, we conclude that it will occur with probability $\frac{1}{2}$, giving $P(Y_2 = 3) = \frac{1}{2}$. We can similarly compute the probability of any counterfactual statement, for example, $P(Y_1 = 4) = \frac{1}{6}, P(Y_1 = 3) = \frac{1}{3}, P(Y_2 > 3) = \frac{1}{2}$, and so on. What is remarkable, however, is that we can also compute joint probabilities of every combination of counterfactual and observable events. For example,

$$P(Y_2 > 3, Y_1 < 4) = \frac{1}{3}$$

$$P(Y_1 < 4, Y - X > 1) = \frac{1}{3}$$

$$P(Y_1 < Y_2) = 1$$

In the first of these expressions, we find a joint probability of two events occurring in two different worlds; the first $Y_2 > 3$ in an $X = 2$ world, and the second $Y_1 < 4$, in $X = 1$. The probability of their conjunction evaluates to $\frac{1}{3}$ because the two events co-occur only at $U = 2$, which was assigned a probability of $\frac{1}{3}$. Other cross-world events appear in the second and third expressions. Remarkably (and usefully), this clash between the worlds provides no barrier to calculation. In fact, cross-world probabilities are as simple to derive as intra-world ones: We simply identify the rows in which the specified combination is true and sum up the probabilities assigned to those rows. This immediately gives us the capability of computing conditional probabilities among counterfactuals and defining notions such as dependence and conditional independence among counterfactuals, as we did in Chapter 1 when we dealt with observable variables. For instance, it is easy to verify that, among individuals for which Y is greater than 2, the probability is $\frac{2}{3}$ that Y would increase if X were 3. (Because $P(Y_3 > Y|Y > 2) = \frac{1}{3}/\frac{1}{2} = \frac{2}{3}$.) Similarly, we can verify that the difference $Y_{x+1} - Y_x$ is independent of x, which means that the causal effect of X on Y does not vary across population types, a property shared by all linear models.

Such joint probabilities over multiple-world counterfactuals can easily be expressed using the subscript notation, as in $P(Y_1 = y_1, Y_2 = y_2)$, and can be computed from any structural model as we did in Table 4.1. They cannot however be expressed using the $do(x)$ notation, because the latter delivers just one probability for each intervention $X = x$. To see the ramifications of this limitation, let us examine a slight modification of the model in Eqs. (4.3) and (4.4), in which a third variable Z acts as mediator between X and Y. The new model's equations are given by

$$X = U_1 \quad Z = aX + U_2, Y = bZ \tag{4.7}$$

and its structure is depicted in Figure 4.3. To cast this model in a context, let $X = 1$ stand for having a college education, $U_2 = 1$ for having professional experience, Z for the level of skill needed for a given job, and Y for salary.

Suppose our aim is to compute $E[Y_{X=1}|Z = 1]$, which stands for the expected salary of individuals with skill level $Z = 1$, had they received a college education. This quantity cannot be captured by a *do*-expression, because the condition $Z = 1$ and the antecedent $X = 1$ refer to two different worlds; the former represents current skills, whereas the latter represents a hypothetical education in an unrealized past. An attempt to capture this hypothetical salary using the expression $E[Y|do(X = 1), Z = 1]$ would not reveal the desired information. The *do*-expression stands for the expected salary of individuals who all finished college and have since acquired skill level $Z = 1$. The salaries of these individuals, as the graph shows, depend only on their skill, and are not affected by whether they obtained the skill through college or through work experience. Conditioning on $Z = 1$, in this case, cuts off the effect of the intervention that we're interested in. In contrast, some of those who currently have $Z = 1$ might not have gone to college and would have attained higher skill (and salary) had they gotten college education. Their salaries are of great interest to us, but they are not included in the *do*-expression. Thus, in general, the *do*-expression will not capture our counterfactual question:

$$E[Y|do(X = 1),\ Z = 1] \neq E[Y_{X=1}|Z = 1] \tag{4.8}$$

We can further confirm this inequality by noting that, while $E[Y|do(X = 1), Z = 1]$ is equal to $E[Y|do(X = 0), Z = 1], E[Y_{X=1}|Z = 1]$ is not equal to $E[Y_{X=0}|Z = 1]$; the formers treat $Z = 1$ as a postintervention condition that prevails for two different sets of units under the two antecedents, whereas the latters treat it as defining *one* set of units in the current world that would react differently under the two antecedents. The $do(x)$ notation cannot capture the latters because the events $X = 1$ and $Z = 1$ in the expression $E[Y_{X=1}|Z = 1]$ refer to two different worlds, pre- and postintervention, respectively. The expression $E[Y|do(X = 1), Z = 1]$

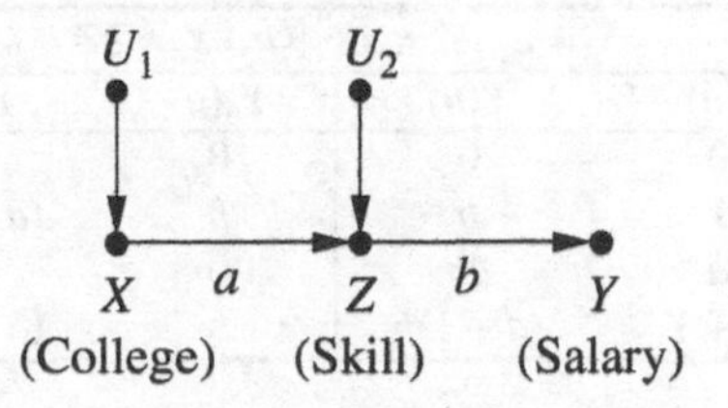

Figure 4.3 A model representing Eq. (4.7), illustrating the causal relations between college education (X), skills (Z), and salary (Y)

on the other hand, invokes only postintervention events, and that is why it is expressible in $do(x)$ notation.

A natural question to ask is whether counterfactual notation can capture the postintervention, single-world expression $E[Y|do(X=1), Z=1]$. The answer is affirmative; being more flexible, counterfactuals can capture both single-world and cross-world probabilities. The translation of $E[Y|do(X=1), Z=1]$ into counterfactual notation is simply $E[Y_{X=1}|Z_{X=1}=1]$, which explicitly designates the event $Z=1$ as postintervention. The variable $Z_{X=1}$ stands for the value that Z would attain had X been 1, and this is precisely what we mean when we put $Z=z$ in a *do*-expression by Bayes' rule:

$$P(Y=y|do(X=1), Z=z) = \frac{P(Y=y, Z=z|do(X=1))}{P(Z=z|do(X=1))}$$

This shows explicitly how the dependence of Z on X should be treated. In the special case where Z is a preintervention variable, as age was in our discussion of conditional interventions (Section 3.5) we have $Z_{X=1}=Z$, and we need not distinguish between the two. The inequality in (4.8) then turns into an equality.

Let's look at how this logic is reflected in the numbers. Table 4.2 depicts the counterfactuals associated with the model of (4.7), with all subscripts denoting the state of X. It was constructed by the same method we used in constructing Table 4.1: replacing the equation $X=u$ with the appropriate constant (zero or one) and solving for Y and Z. Using this table, we can verify immediately that (see footnote 2)

$$E[Y_1|Z=1] = (a+1)b \tag{4.9}$$

$$E[Y_0|Z=1] = b \tag{4.10}$$

$$E[Y|do(X=1), Z=1] = b \tag{4.11}$$

$$E[Y|do(X=0), Z=1] = b \tag{4.12}$$

These equations provide numerical confirmation of the inequality in (4.8). They also demonstrate a peculiar property of counterfactual conditioning that we have noted before: Despite the fact that Z separates X from Y in the graph of Figure 4.3, we find that X has an effect on Y for those units falling under $Z=1$:

$$E[Y_1 - Y_0|Z=1] = ab \neq 0$$

The reason for this behavior is best explained in the context of our salary example. While the salary of those who have acquired skill level $Z=1$ depends only on their skill, not on X, the

Table 4.2 The values attained by $X(u), Y(u), Z(u), Y_0(u), Y_1(u), Z_0(u)$, and $Z_1(u)$ in the model of Eq. (4.7)

				$X=u_1$ $Z=aX+u_2$ $Y=bZ$				
u_1	u_2	$X(u)$	$Z(u)$	$Y(u)$	$Y_0(u)$	$Y_1(u)$	$Z_0(u)$	$Z_1(u)$
0	0	0	0	0	0	ab	0	a
0	1	0	1	b	b	$(a+1)b$	1	$a+1$
1	0	1	a	ab	0	ab	0	a
1	1	1	$a+1$	$(a+1)b$	b	$(a+1)b$	1	$a+1$

[2] Strictly speaking, the quantity $E[Y|\ do(X=1), Z=1]$ in Eq. (4.11) is undefined because the observation $Z=1$ is not possible post-intervention of $do(X=1)$. However, for the purposes of the example, we can imagine that $Z=1$ was observed due to some error term $\epsilon_z \rightarrow Z$ that accounts for the deviation. Eq. (4.11) then follows.

salary of those who are currently at $Z = 1$ would have been different *had they had a different past*. Retrospective reasoning of this sort, concerning dependence on the unrealized past, is not shown explicitly in the graph of Figure 4.3. To facilitate such reasoning, we need to devise means of representing counterfactual variables directly in the graph; we provide such representations in Section 4.3.2.

Thus far, the relative magnitudes of the probabilities of $P(u_1)$ and $P(u_2)$ have not entered into the calculations, because the condition $Z = 1$ occurs only for $u_1 = 0$ and $u_2 = 1$ (assuming that $a \neq 0$ and $a \neq 1$), and under these conditions, each of Y, Y_1, and Y_0 has a definite value. These probabilities play a role, however, if we assume $a = 1$ in the model, since $Z = 1$ can now occur under two conditions: $(u_1 = 0, u_2 = 1)$ and $(u_1 = 1, u_2 = 0)$. The first occurs with probability $P(u_1 = 0)P(u_2 = 1)$ and the second with probability $P(u_1 = 1)P(u_2 = 0)$. In such a case, we obtain

$$E[Y_{X=1}|Z = 1] = b\left(1 + \frac{P(u_1 = 0)P(u_2 = 1)}{P(u_1 = 0)P(u_2 = 1) + P(u_1 = 1)P(u_2 = 0)}\right) \tag{4.13}$$

$$E[Y_{X=0}|Z = 1] = b\left(\frac{P(u_1 = 0)P(u_2 = 1)}{P(u_1 = 0)P(u_2 = 1) + P(u_1 = 1)P(u_2 = 0)}\right) \tag{4.14}$$

The fact that the first expression is larger than the second demonstrates again that the skill-specific causal effect of education on salary is nonzero, despite the fact that salaries are determined by skill only, not by education. This is to be expected, since a nonzero fraction of the workers at skill level $Z = 1$ did not receive college education, and, had they been given college education, their skill would have increased to $Z_1 = 2$, and their salaries to $2b$.

Study question 4.3.1

Consider the model in Figure 4.3 and assume that U_1 and U_2 are two independent Gaussian variables, each with zero mean and unit variance.

(a) *Find the expected salary of workers at skill level $Z = z$ had they received x years of college education. [Hint: Use Theorem 4.3.2, with $e : Z = z$, and the fact that for any two Gaussian variables, say X and Z, we have $E[X|Z = z] = E[X] + R_{XZ}\,(z - E[Z])$. Use the material in Sections 3.8.2 and 3.8.3 to express all regression coefficients in terms of structural parameters, and show that $E[Y_x|Z = z] = abx + bz/(1 + a^2)$.]*

(b) *Based on the solution for (a), show that the skill-specific effect of education on salary is independent of the skill level.*

4.3.2 *The Graphical Representation of Counterfactuals*

Since counterfactuals are byproducts of structural equation models, a natural question to ask is whether we can see them in the causal graphs associated with those models. The answer is affirmative, as can be seen from the fundamental law of counterfactuals, Eq. (4.5). This law tells us that if we modify model M to obtain the submodel M_x, then the outcome variable Y in the modified model is the counterfactual Y_x of the original model. Since modification calls for removing all arrows entering the variable X, as illustrated in Figure 4.4, we conclude that the node associated with the Y variable serves as a surrogate for Y_x, with the understanding that the substitution is valid only under the modification.

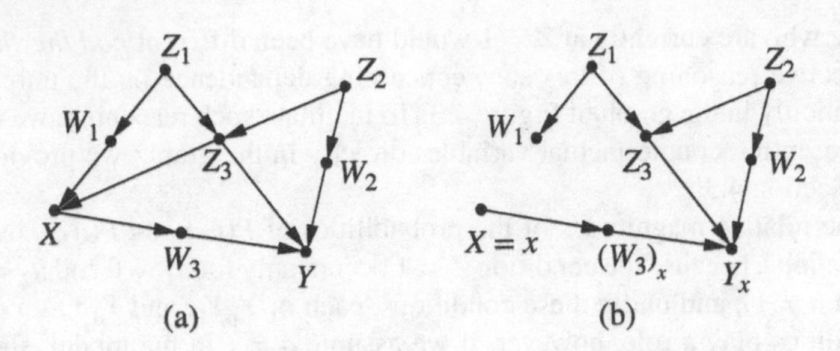

Figure 4.4 Illustrating the graphical reading of counterfactuals. (a) The original model. (b) The modified model M_x in which the node labeled Y_x represents the potential outcome Y predicated on $X = x$

This temporary visualization of counterfactuals is sufficient to answer some fundamental questions about the statistical properties of Y_x and how those properties depend on other variables in the model, specifically when those other variables are conditioned on.

When we ask about the statistical properties of Y_x, we need to examine what would cause Y_x to vary. According to its structural definition, Y_x represents the value of Y under a condition where X is held constant at $X = x$. Statistical variations of Y_x are therefore governed by all exogenous variables capable of influencing Y when X is held constant, that is, when the arrows entering X are removed, as in Figure 4.4(b). Under such conditions, the set of variables capable of transmitting variations to Y are the parents of Y (observed and unobserved), as well as parents of nodes on the pathways between X and Y. In Figure 4.4(b), for example, these parents are $\{Z_3, W_2, U_3, U_Y\}$, where U_Y and U_3, the error terms of Y and W_3, are not shown in the diagram. (These variables remain the same in both models.) Any set of variables that blocks a path to these parents also blocks that path to Y_x, and will result in, therefore, a conditional independence for Y_x. In particular, if we have a set Z of covariates that satisfies the backdoor criterion in M (see Definition 3.3.1), that set also blocks all paths between X and those parents, and consequently, it renders X and Y_x independent in every stratum $Z = z$.

These considerations are summarized formally in Theorem 4.3.1.

Theorem 4.3.1 (Counterfactual Interpretation of Backdoor) *If a set Z of variables satisfies the backdoor condition relative to (X, Y), then, for all x, the counterfactual Y_x is conditionally independent of X given Z*

$$P(Y_x|X, Z) = P(Y_x|Z) \tag{4.15}$$

Theorem 4.3.1 has far-reaching consequences when it comes to estimating the probabilities of counterfactuals from observational studies. In particular, it implies that $P(Y_x = y)$ is identifiable by the adjustment formula of Eq. (3.5). To prove this, we condition on Z (as in Eq. (1.9)) and write

$$\begin{aligned} P(Y_x = y) &= \sum_z P(Y_x = y|Z = z)P(z) \\ &= \sum_z P(Y_x = y|Z = z, X = x)P(z) \\ &= \sum_z P(Y = y|Z = z, X = x)P(z) \end{aligned} \tag{4.16}$$

The second line was licensed by Theorem 4.3.1, whereas the third line was licensed by the consistency rule (4.6).

The fact that we obtained the familiar adjustment formula in Eq. (4.16) is not really surprising, because this same formula was derived in Section 3.2 (Eq. (3.4)), for $P(Y = y|do(x))$, and we know that $P(Y_x = y)$ is just another way of writing $P(Y = y|do(x))$. Interestingly, this derivation invokes only algebraic steps; it makes no reference to the model once we ensure that Z satisfies the backdoor criterion. Equation (4.15), which converts this graphical reality into algebraic notation, and allows us to derive (4.16), is sometimes called "conditional ignorability"; Theorem 4.3.1 gives this notion a scientific interpretation and permits us to test whether it holds in any given model.

Having a graphical representation for counterfactuals, we can resolve the dilemma we faced in Section 4.3.1 (Figure 4.3), and explain graphically why a stronger education (X) would have had an effect on the salary (Y) of people who are currently at skill level $Z = z$, despite the fact that, according to the model, salary is determined by skill only. Formally, to determine if the effect of education on salary (Y_x) is statistically independent of the level of education, we need to locate Y_x in the graph and see if it is d-separated from X given Z. Referring to Figure 4.3, we see that Y_x can be identified with U_2, the only parent of nodes on the causal path from X to Y (and therefore, the only variable that produces variations in Y_x while X is held constant). A quick inspection of Figure 4.3 tells us that Z acts as a collider between X and U_2, and, therefore, X and U_2 (and similarly X and Y_x) are not d-separated given Z. We conclude therefore

$$E[Y_x|X, Z] \neq E[Y_x|Z]$$

despite the fact that

$$E[Y|X, Z] = E[Y|Z]$$

In Study question 4.3.1, we evaluate these counterfactual expectations explicitly, assuming a linear Gaussian model. The graphical representation established in this section permits us to determine independencies among counterfactuals by graphical means, without assuming linearity or any specific parametric form. This is one of the tools that modern causal analysis has introduced to statistics, and, as we have seen in the analysis of the education–skill–salary story, it takes a task that is extremely hard to solve by unaided intuition and reduces it to simple operations on graphs. Additional methods of visualizing counterfactual dependencies, called "twin networks," are discussed in (Pearl 2000, pp. 213–215).

4.3.3 *Counterfactuals in Experimental Settings*

Having convinced ourselves that every counterfactual question can be answered from a fully specified structural model, we next move to the experimental setting, where a model is not available, and the experimenter must answer interventional questions on the basis of a finite sample of observed individuals. Let us refer back to the "encouragement design" model of Figure 4.1, in which we analyzed the behavior of an individual named Joe, and assume that the experimenter observes a set of 10 individuals, with Joe being participant 1. Each individual is characterized by a distinct vector $U_i = (U_X, U_H, U_Y)$, as shown in the first three columns of Table 4.3.

Table 4.3 Potential and observed outcomes predicted by the structural model of Figure 4.1 units were selected at random, with each U_i uniformly distributed over [0, 1]

	Participant characteristics			Observed behavior			Predicted potential outcomes				
Participant	U_X	U_H	U_Y	X	Y	H	Y_0	Y_1	H_0	H_1	$Y_{00} \cdots$
1	0.5	0.75	0.75	0.5	1.50	1.0	1.05	1.95	0.75	1.25	0.75
2	0.3	0.1	0.4	0.3	0.71	0.25	0.44	1.34	0.1	0.6	0.4
3	0.5	0.9	0.2	0.5	1.01	1.15	0.56	1.46	0.9	1.4	0.2
4	0.6	0.5	0.3	0.6	1.04	0.8	0.50	1.40	0.5	1.0	0.3
5	0.5	0.8	0.9	0.5	1.67	1.05	1.22	2.12	0.8	1.3	0.9
6	0.7	0.9	0.3	0.7	1.29	1.25	0.66	1.56	0.9	1.4	0.3
7	0.2	0.3	0.8	0.2	1.10	0.4	0.92	1.82	0.3	0.8	0.8
8	0.4	0.6	0.2	0.4	0.80	0.8	0.44	1.34	0.6	1.1	0.2
9	0.6	0.4	0.3	0.6	1.00	0.7	0.46	1.36	0.4	0.9	0.3
10	0.3	0.8	0.3	0.3	0.89	0.95	0.62	1.52	0.8	1.3	0.3

Using this information, we can create a full data set that complies with the model. For each triplet (U_X, U_H, U_Y), the model of Figure 4.1 enables us to complete a full row of the table, including Y_0 and Y_1, which stand for the potential outcomes under treatment $(X = 1)$ and control $(X = 0)$ conditions, respectively. We see that the structural model in Figure 4.1 encodes in effect a synthetic population of individuals together with their predicted behavior under both observational and experimental conditions. The columns labeled X, Y, H predict the results of observational studies, and those labeled Y_0, Y_1, H_0, H_1 predict the hypothetical outcome under two treatment regimes, $X = 0$, and $X = 1$. Many more, in fact infinite, potential outcomes may be predicted; for example, $Y_{X=0.5,H=2.0}$ as computed for Joe from Figure 4.2, as well as all combinations of subscripted variables. From this synthetic population, one can estimate the probability of every counterfactual query on variables X, Y, H, assuming, of course, that we are in possession of all entries of the table. The estimation would require us to simply count the proportion of individuals that satisfy the specified query as demonstrated in Section 4.3.1.

Needless to say, the information conveyed by Table 4.3 is not available to us in either observational or experimental studies. This information was deduced from a parametric model such as the one in Figure 4.2, from which we could infer the defining characteristics $\{U_X, U_H, U_Y\}$ of each participant, given the observations $\{X, H, Y\}$. In general, in the absence of a parametric model, there is very little we learn about the potential outcomes Y_1 and Y_0 of individual participants, when all we have is their observed behavior $\{X, H, Y\}$. Theoretically, the only connection we have between the counterfactuals $\{Y_1, Y_0\}$ and the observables $\{X, H, Y\}$ is the consistency rule of Eq. (4.6), which informs us that Y_1 must be equal to Y in case $X = 1$ and Y_0 must be equal to Y in case $X = 0$. But aside from this tenuous connection, most of the counterfactuals associated with the individual participants will remain unobserved.

Fortunately, there is much we can learn about those counterfactuals at the population level, such as estimating their probabilities or expectation. This we have witnessed already through the adjustment formula of (4.16), where we were able to compute $E(Y_1 - Y_0)$ using the graph alone, instead of a complete model. Much more can be obtained from experimental studies, where even the graph becomes dispensable.

Assume that we have no information whatsoever about the underlying model. All we have are measurements on Y taken in an experimental study in which X is randomized over two levels, $X = 0$ and $X = 1$.

Table 4.4 describes the responses of the same 10 participants (Joe being participant 1) under such experimental conditions, with participants 1, 5, 6, 8, and 10 assigned to $X = 0$, and the rest to $X = 1$. The first two columns give the true potential outcomes (taken from Table 4.3), while the last two columns describe the information available to the experimenter, where a square indicates that the response was not observed. Clearly, Y_0 is observed only for participants assigned to $X = 0$ and, similarly, Y_1 is observed only for those assigned to $X = 1$. Randomization assures us that, although half of the potential outcomes are not observed, the difference between the observed *means* in the treatment and control groups will converge to the difference of the population averages, $E(Y_1 - Y_0) = 0.9$. This is because randomization distributes the black squares at random along the two rightmost columns of Table 4.4, independent of the actual values of Y_0 and Y_1, so as the number of samples increases, the sample means converge to the population means.

This unique and important property of randomized experiments is not new to us, since randomization, like interventions, renders X independent of any variable that may affect Y (as in Figure 4.4(b)). Under such conditions, the adjustment formula (4.16) is applicable with $Z = \{\ \}$, yielding $E[Y_x] = E[Y|X = x]$, where $x = 1$ represents treated units and $x = 0$ untreated. Table 4.4 helps us understand what is actually computed when we take sample averages in experimental settings and how those averages are related to the underlying counterfactuals, Y_1 and Y_0.

Table 4.4 Potential and observed outcomes in a randomized clinical trial with X randomized over $X = 0$ and $X = 1$

	Predicted potential outcomes		Observed outcomes	
Participant	Y_0	Y_1	Y_0	Y_1
1	1.05	1.95	1.05	■
2	0.44	1.34	■	1.34
3	0.56	1.46	■	1.46
4	0.50	1.40	■	1.40
5	1.22	2.12	1.22	■
6	0.66	1.56	0.66	■
7	0.92	1.82	■	1.82
8	0.44	1.34	0.44	■
9	0.46	1.36	■	1.36
10	0.62	1.52	0.62	■
	True average treatment effect: 0.90		Study average treatment effect: 0.68	

4.3.4 Counterfactuals in Linear Models

In nonparametric models, counterfactual quantities of the form $E[Y_{X=x}|Z = z]$ may not be identifiable, even if we have the luxury of running experiments. In fully linear models, however, things are much easier; any counterfactual quantity is identifiable whenever the model parameters are identified. This is because the parameters fully define the model's functions, and as we have seen earlier, once the functions are given, counterfactuals are computable using Eq. (4.5). Since every model parameter is identifiable from interventional studies using the interventional definition of direct effects, we conclude that in linear models, every counterfactual is experimentally identifiable. The question remains whether counterfactuals can be identified in observational studies, when some of the model parameters are not identified. It turns out that any counterfactual of the form $E[Y_{X=x}|Z = e]$, with e an arbitrary set of evidence, is identified whenever $E[Y|do(X = x)]$ is identified (Pearl 2000, p. 389). The relation between the two is summarized in Theorem 4.3.2, which provides a shortcut for computing counterfactuals.

Theorem 4.3.2 *Let τ be the slope of the total effect of X on Y,*

$$\tau = E[Y|do(x + 1)] - E[Y|do(x)]$$

then, for any evidence $Z = e$, we have

$$E[Y_{X=x}|Z = e] = E[Y|Z = e] + \tau(x - E[X|Z = e]) \tag{4.17}$$

This provides an intuitive interpretation of counterfactuals in linear models: $E[Y_{X=x}|Z = e]$ can be computed by first calculating the best estimate of Y conditioned on the evidence e, $E[Y|e]$, and then adding to it whatever change is expected in Y when X is shifted from its current best estimate, $E[X|Z = e]$, to its hypothetical value, x.

Methodologically, the importance of Theorem 4.3.2 lies in enabling researchers to answer hypothetical questions about individuals (or sets of individuals) from population data. The ramifications of this feature in legal and social contexts will be explored in the following sections. In the situation illustrated by Figure 4.2, we computed the counterfactual $Y_{H=2}$ under the evidence $e = \{X = 0.5, H = 1, Y = 1\}$. We now demonstrate how Theorem 4.3.2 can be applied to this model in computing the *effect of treatment on the treated*

$$ETT = E[Y_1 - Y_0|X = 1] \tag{4.18}$$

Substituting the evidence $e = \{X = 1\}$ in Eq. (4.17) we get

$$\begin{aligned} ETT &= E[Y_1|X = 1] - E[Y_0|X = 1] \\ &= E[Y|X = 1] - E[Y|X = 1] + \tau(1 - E[X|X = 1]) - \tau(0 - E[X|X = 1]) \\ &= \tau \\ &= b + ac = 0.9 \end{aligned}$$

In other words, the effect of treatment on the treated is equal to the effect of treatment on the entire population. This is a general result in linear systems that can be seen directly from Eq. (4.17); $E[Y_{x+1} - Y_x|e] = \tau$, independent on the evidence of e. Things are different when a multiplicative (i.e., nonlinear) interaction term is added to the output equation. For example, if the arrow $X \rightarrow H$ were reversed in Figure 4.1, and the equation for Y read

$$Y = bX + cH + \delta XH + U_Y \tag{4.19}$$

τ would differ from ETT. We leave it to the reader as an exercise to show that the difference $ETT - \tau$ then equals $\frac{\delta a}{1+a^2}$ (see Study question 4.3.2(c)).

Study questions

Study question 4.3.2

(a) Describe how the parameters a, b, c in Figure 4.1 can be estimated from nonexperimental data.

(b) In the model of Figure 4.3, find the effect of education on those students whose salary is $Y = 1$. [Hint: Use Theorem 4.3.2 to compute $E[Y_1 - Y_0|Y = 1]$.]

(c) Estimate τ and the $ETT = E[Y_1 - Y_0|X = 1]$ for the model described in Eq. (4.19). [Hint: Use the basic definition of counterfactuals, Eq. (4.5) and the equality $E[Z|X = x'] = R_{ZX}x'$.]

4.4 Practical Uses of Counterfactuals

Now that we know how to compute counterfactuals, it will be instructive—and motivating—to see counterfactuals put to real use. In this section, we examine examples of problems that seem baffling at first, but that can be solved using the techniques we just laid out. Hopefully, the reader will leave this chapter with both a better understanding of how counterfactuals are used and a deeper appreciation of why we would want to use them.

4.4.1 Recruitment to a Program

Example 4.4.1 *A government is funding a job training program aimed at getting jobless people back into the workforce. A pilot randomized experiment shows that the program is effective; a higher percentage of people were hired among those who finished the program than among those who did not go through the program. As a result, the program is approved, and a recruitment effort is launched to encourage enrollment among the unemployed, by offering the job training program to any unemployed person who elects to enroll.*

Lo and behold, enrollment is successful, and the hiring rate among the program's graduates turns out even higher than in the randomized pilot study. The program developers are happy with the results and decide to request additional funding.

Oddly, critics claim that the program is a waste of taxpayers' money and should be terminated. Their reasoning is as follows: While the program was somewhat successful in the experimental study, where people were chosen at random, there is no proof that the program accomplishes its mission among those who choose to enroll of their own volition. Those who self-enroll, the critics say, are more intelligent, more resourceful, and more socially connected than the eligible who did not enroll, and are more likely to have found a job regardless of the training. The critics claim that what we need to estimate is the *differential* benefit of the program on those enrolled: the extent to which hiring rate has increased among the enrolled, compared to what it would have been had they not been trained.

Using our subscript notation for counterfactuals, and letting $X = 1$ represent training and $Y = 1$ represent hiring, the quantity that needs to be evaluated is the effect of training on the trained (ETT, better known as "effect of treatment on the treated," Eq. (4.18)):

$$ETT = E[Y_1 - Y_0|X = 1] \tag{4.20}$$

Here the difference $Y_1 - Y_0$ represents the causal effect of training (X) on hiring (Y) for a randomly chosen individual, and the condition $X = 1$ limits the choice to those actually choosing the training program on their own initiative.

As in our freeway example of Section 4.1, we are witnessing a clash between the antecedent ($X = 0$) of the counterfactual Y_0 (hiring had training not taken place) and the event it is conditioned on, $X = 1$. However, whereas the counterfactual analysis in the freeway example had no tangible consequences save for a personal regret statement—"I should have taken the freeway"—here the consequences have serious economic implications, such as terminating a training program, or possibly restructuring the recruitment strategy to attract people who would benefit more from the program offered.

The expression for ETT does not appear to be estimable from either observational or experimental data. The reason rests, again, in the clash between the subscript of Y_0 and the event $X = 1$ on which it is conditioned. Indeed, $E[Y_0|X = 1]$ stands for the expectation that a trained person ($X = 1$) would find a job had he/she not been trained. This counterfactual expectation seems to defy empirical measurement because we can never rerun history and deny training to those who received it. However, we see in the subsequent sections of this chapter that, despite this clash of worlds, the expectation $E[Y_0|X = 1]$ can be reduced to estimable expressions in many, though not all, situations. One such situation occurs when a set Z of covariates satisfies the backdoor criterion with regard to the treatment and outcome variables. In such a case, ETT probabilities are given by a modified adjustment formula:

$$\begin{aligned} P(Y_x = y|X = x') \\ = \sum_z P(Y = y|X = x, Z = z)P(Z = z|X = x') \end{aligned} \tag{4.21}$$

This follows directly from Theorem 4.3.1, since conditioning on $Z = z$ gives

$$P(Y_x = y|x') = \sum_z P(Y_x = y|z, x')P(z|x')$$

but Theorem 4.3.1 permits us to replace x' with x, which by virtue of (4.6) permits us to remove the subscript x from Y_x, yielding (4.21).

Comparing (4.21) to the standard adjustment formula of Eq. (3.5),

$$P(Y = y|do(X = x)) = \sum P(Y = y|X = x, Z = z)P(Z = z)$$

we see that both formulas call for conditioning on $Z = z$ and averaging over z, except that (4.21) calls for a different weighted average, with $P(Z = z|X = x')$ replacing $P(Z = z)$.

Using Eq. (4.21), we readily get an estimable, noncounterfactual expression for ETT

$$
\begin{aligned}
ETT &= E[Y_1 - Y_0|X = 1] \\
&= E[Y_1|X = 1] - E[Y_0|X = 1] \\
&= E[Y|X = 1] - \sum_z E[Y|X = 0, Z = z]P(Z = z|X = 1)
\end{aligned}
$$

where the first term in the final expression is obtained using the consistency rule of Eq. (4.6). In other words, $E[Y_1|X = 1] = E[Y|X = 1]$ because, conditional on $X = 1$, the value that Y would get had X been 1 is simply the observed value of Y.

Another situation permitting the identification of ETT occurs for binary X whenever both experimental and nonexperimental data are available, in the form of $P(Y = y|do(X = x))$ and $P(X = x, Y = y)$, respectively. Still another occurs when an intermediate variable is available between X and Y satisfying the front-door criterion (Figure 3.10(b)). What is common to these situations is that an inspection of the causal graph can tell us whether ETT is estimable and, if so, how.

Study questions

Study question 4.4.1

(a) *Prove that, if X is binary, the effect of treatment on the treated can be estimated from both observational and experimental data. Hint: Decompose* $E[Y_x]$ *into*

$$E[Y_x] = E[Y_x|x']P(x') + E[Y_x|x]P(x)$$

(b) *Apply the result of Question (a) to Simpson's story with the nonexperimental data of Table 1.1, and estimate the effect of treatment on those who used the drug by choice. [Hint: Estimate* $E[Y_x]$ *assuming that gender is the only confounder.]*

(c) *Repeat Question (b) using Theorem 4.3.1 and the fact that Z in Figure 3.3 satisfies the backdoor criterion. Show that the answers to (b) and (c) coincide.*

4.4.2 Additive Interventions

Example 4.4.2 *In many experiments, the external manipulation consists of adding (or subtracting) some amount from a variable X without disabling preexisting causes of X, as required by the do(x) operator. For example, we might give 5 mg/l of insulin to a group of patients with varying levels of insulin already in their systems. Here, the preexisting causes of the manipulated variable continue to exert their influences, and a new quantity is added, allowing for differences among units to continue. Can the effect of such interventions be predicted from observational studies, or from experimental studies in which X was set uniformly to some predetermined value x?*

If we write our question using counterfactual variables, the answer becomes obvious. Suppose we were to add a quantity q to a treatment variable X that is currently at level $X = x'$.

The resulting outcome would be $Y_{x'+q}$, and the average value of this outcome over all units currently at level $X = x'$ would be $E[Y_x|x']$, with $x = x' + q$. Here, we meet again the ETT expression $E[Y_x|x']$, to which we can apply the results described in the previous example. In particular, we can conclude immediately that, whenever a set Z in our model satisfies the backdoor criterion, the effect of an additive intervention is estimable using the ETT adjustment formula of Eq. (4.21). Substituting $x = x' + q$ in (4.21) and taking expectations gives the effect of this intervention, which we call *add*(q):

$$
\begin{aligned}
&E[Y|add(q)] - E[Y] \\
&= \sum_{x'} E[Y_{x'+q}|X = x']P(X = x') - E[Y] \\
&= \sum_{x'} \sum_{z} E[Y|X = x' + q, Z = z]P(Z = z|X = x')P(X = x') - E[Y] \qquad (4.22)
\end{aligned}
$$

In our example, Z may include variables such as age, weight, or genetic disposition; we require only that each of those variables be measured and that they satisfy the backdoor condition.

Similarly, estimability is assured for all other cases in which ETT is identifiable.

This example demonstrates the use of counterfactuals to estimate the effect of practical interventions, which cannot always be described as *do*-expressions, but may nevertheless be estimated under certain circumstances. A question naturally arises: Why do we need to resort to counterfactuals to predict the effect of a rather common intervention, one that could be estimated by a straightforward clinical trial at the population level? We simply split a randomly chosen group of subjects into two parts, subject half of them to an *add*(q) type of intervention and compare the expected value of Y in this group to that obtained in the *add*(0) group. What is it about additive interventions that force us to seek the advice of a convoluted oracle, in the form of counterfactuals and ETT, when the answer can be obtained by a simple randomized trial?

The answer is that we need to resort to counterfactuals only because our target quantity, $E[Y|add(q)]$, could not be reduced to *do*-expressions, and it is through *do*-expressions that scientists report their experimental findings. This does not mean that the desired quantity $E[Y|add(q)]$ cannot be obtained from a specially designed experiment; it means only that save for conducting such a special experiment, the desired quantity cannot be inferred from scientific knowledge or from a standard experiment in which X is set to $X = x$ uniformly over the population. The reason we seek to base policies on such ideal standard experiments is that they capture scientific knowledge. Scientists are interested in quantifying the effect of increasing insulin concentration in the blood from a given level $X = x$ to a another level $X = x + q$, and this increase is captured by the *do*-expression: $E[Y|do(X = x + q)] - E[Y|do(X = x)]$. We label it "scientific" because it is biologically meaningful, namely its implications are invariant across populations (indeed laboratory blood tests report patients' concentration levels, $X = x$, which are tracked over time). In contrast, the policy question in the case of additive interventions does not have this invariance feature; it asks for the average effect of adding an increment q to everyone, regardless of the current x level of each individual in this particular population. It is not immediately transportable, because it is highly sensitive to the probability $P(X = x)$ in the population under study. This creates a mismatch between what science tells us and what policy makers ask us to estimate. It is no wonder, therefore, that we need to resort to a unit-level analysis (i.e., counterfactuals) in order to translate from one language into another.

The reader may also wonder why $E[Y|add(q)]$ is not equal to the average causal effect

$$\sum_x [E[Y|do(X = x + q)] - E[Y|do(X = x)]] \; P(X = x)$$

After all, if we know that adding q to an individual at level $X = x$ would increase its expected Y by $E[Y|do(X = x + q)] - E[Y|do(X = x)]$, then averaging this increase over X should give us the answer to the policy question $E[Y|add(q)]$. Unfortunately, this average does *not* capture the policy question. This average represents an experiment in which subjects are chosen at random from the population, a fraction $P(X = x)$ are given an additional dose q, and the rest are left alone. But things are different in the policy question at hand, since $P(X = x)$ represents the proportion of subjects who entered level $X = x$ by free choice, and we cannot rule out the possibility that subjects who attain $X = x$ by free choice would react to $add(q)$ differently from subjects who "receive" $X = x$ by experimental decree. For example, it is quite possible that subjects who are highly sensitive to $add(q)$ would attempt to lower their X level, given the choice.

We translate into counterfactual analysis and write the inequality:

$$E[Y|add(q)] = \sum_x E[Y_{x+q}|x]P(X = x) \neq \sum_x E[Y_{x+q}]P(X = x)$$

Equality holds only when Y_x is independent of X, a condition that amounts to nonconfounding (see Theorem 4.3.1). Absent this condition, the estimation of $E[Y|add(q)]$ can be accomplished either by q-specific intervention or through stronger assumptions that enable the translation of ETT to *do*-expressions, as in Eq. (4.21).

Study question 4.4.2

Joe has never smoked before but, as a result of peer pressure and other personal factors, he decided to start smoking. He buys a pack of cigarettes, comes home, and asks himself: "I am about to start smoking, should I?"

(a) *Formulate Joe's question mathematically, in terms of ETT, assuming that the outcome of interest is lung cancer.*
(b) *What type of data would enable Joe to estimate his chances of getting cancer given that he goes ahead with the decision to smoke, versus refraining from smoking.*
(c) *Use the data in Table 3.1 to estimate the chances associated with the decision in (b).*

4.4.3 Personal Decision Making

Example 4.4.3 *Ms Jones, a cancer patient, is facing a tough decision between two possible treatments: (i) lumpectomy alone or (ii) lumpectomy plus irradiation. In consultation with her oncologist, she decides on (ii). Ten years later, Ms Jones is alive, and the tumor has not recurred. She speculates: Do I owe my life to irradiation?*

Mrs Smith, on the other hand, had a lumpectomy alone, and her tumor recurred after a year. And she is regretting: I should have gone through irradiation.

Can these speculations ever be substantiated from statistical data? Moreover, what good would it do to confirm Ms Jones's triumph or Mrs Smith's regret?

The overall effectiveness of irradiation can, of course, be determined by randomized experiments. Indeed, on October 17, 2002, the *New England Journal of Medicine* published a paper by Fisher et al. describing a 20-year follow-up of a randomized trial comparing lumpectomy alone and lumpectomy plus irradiation. The addition of irradiation to lumpectomy was shown to cause substantially fewer recurrences of breast cancer (14% vs 39%).

These, however, were population results. Can we infer from them the specific cases of Ms Jones and Mrs Smith? And what would we gain if we do, aside from supporting Ms Jones's satisfaction with her decision or intensifying Mrs Smith's sense of failure?

To answer the first question, we must first cast the concerns of Ms Jones and Mrs Smith in mathematical form, using counterfactuals. If we designate remission by $Y = 1$ and the decision to undergo irradiation by $X = 1$, then the probability that determines whether Ms Jones is justified in attributing her remission to the irradiation ($X = 1$) is

$$PN = P(Y_0 = 0|X = 1, Y = 1) \tag{4.23}$$

It reads: the probability that remission would *not* have occurred ($Y = 0$) had Ms Jones not gone through irradiation, given that she did in fact go through irradiation ($X = 1$), and remission did occur ($Y = 1$). The label PN stands for "probability of necessity" that measures the degree to which Ms Jones's decision was *necessary* for her positive outcome.

Similarly, the probability that Mrs Smith's regret is justified is given by

$$PS = P(Y_1 = 1|X = 0, Y = 0) \tag{4.24}$$

It reads: the probability that remission would have occurred had Mrs Smith gone through irradiation ($Y_1 = 1$), given that she did not in fact go through irradiation ($X = 0$), and remission did not occur ($Y = 0$). PS stands for the "probability of sufficiency," measuring the degree to which the action $X = 1$, which was not taken.

We see that these expressions have almost the same form (save for interchanging ones with zeros) and, moreover, both are similar to Eq. (4.1), save for the fact that Y in the freeway example was a continuous variable, so its expected value was the quantity of interest.

These two probabilities (sometimes referred to as "probabilities of causation") play a major role in all questions of "attribution," ranging from legal liability to personal decision making. They are not, in general, estimable from either observational or experimental data, but as we shall see below, they *are* estimable under certain conditions, when both observational and experimental data are available.

But before commencing a quantitative analysis, let us address our second question: What is gained by assessing these retrospective counterfactual parameters? One answer is that notions such as regret and success, being right or being wrong, have more than just emotional value; they play important roles in cognitive development and adaptive learning. Confirmation of Ms Jones's triumph reinforces her confidence in her decision-making strategy, which may include her sources of medical information, her attitude toward risks, and her sense of priority, as well as the strategies she has been using to put all these considerations together. The same applies to regret; it drives us to identify sources of weakness in our strategies and to think of some kind of change that would improve them. It is through counterfactual reinforcement that we learn to improve our own decision-making processes and achieve higher performance. As Kathryn Schultz says in her delightful book *Being Wrong*, "However disorienting, difficult, or humbling our mistakes might be, it is ultimately wrongness, not rightness, that can teach us who we are."

Estimating the probabilities of being right or wrong also has tangible and profound impact on critical decision making. Imagine a third lady, Ms Daily, facing the same decision as Ms Jones did, and telling herself: If my tumor is the type that would not recur under lumpectomy alone, why should I go through the hardships of irradiation? Similarly, if my tumor is the type that would recur regardless of whether I go through irradiation or not, I would rather not go through it. The only reason for me to go through this is if the tumor is the type that would remiss under treatment and recur under no treatment.

Formally, Ms Daily's dilemma is to quantify the probability that irradiation is both *necessary and sufficient* for eliminating her tumor, or

$$PNS = P(Y_1 = 1, Y_0 = 0) \tag{4.25}$$

where Y_1 and Y_0 stand for remission under treatment (Y_1) and nontreatment (Y_0), respectively. Knowing this probability would help Ms Daily's assessment of how likely she is to belong to the group of individuals for whom $Y_1 = 1$ and $Y_0 = 0$.

This probability cannot, of course, be assessed from experimental studies, because we can never tell from experimental data whether an outcome would have been different had the person been assigned to a different treatment. However, casting Ms Daily's question in mathematical form enables us to investigate algebraically what assumptions are needed for estimating PNS and from what type of data. In the next section (Section 4.5.1, Eq. (4.42)), we see that indeed, PNS can be estimated if we assume *monotonicity*, namely, that irradiation cannot cause the recurrence of a tumor that was about to remit. Moreover, under monotonicity, experimental data are sufficient to conclude

$$PNS = P(Y = 1|do(X = 1)) - P(Y = 1|do(X = 0)) \tag{4.26}$$

For example, if we rely on the experimental data of Fisher et al. (2002), this formula permits us to conclude that Ms Daily's PNS is

$$PNS = 0.86 - 0.61 = 0.25$$

This gives her a 25% chance that her tumor is the type that responds to treatment—specifically, that it will remit under lumpectomy plus irradiation but will recur under lumpectomy alone. Such quantification of individual risks is extremely important in personal decision making, and estimates of such risks from population data can only be inferred through counterfactual analysis and appropriate assumptions.

4.4.4 Discrimination in Hiring

Example 4.4.4 *Mary files a law suit against the New York-based XYZ International, alleging discriminatory hiring practices. According to her, she has applied for a job with XYZ International, and she has all the credentials for the job, yet she was not hired, allegedly because she mentioned, during the course of her interview, that she is gay. Moreover, she claims, the hiring record of XYZ International shows consistent preferences for straight employees. Does she have a case? Can hiring records prove whether XYZ International was discriminating when declining her job application?*

At the time of writing, U.S. law doesn't specifically prohibit employment discrimination on the basis of sexual orientation, but New York law does. And New York defines discrimination in much the same way as federal law. U.S. courts have issued clear directives as to what constitutes employment discrimination. According to law makers, "The central question in any employment-discrimination case is whether the employer would have taken the same action had the employee been of a different race (age, sex, religion, national origin, etc.) and everything else had been the same." (In Carson vs Bethlehem Steel Corp., 70 FEP Cases 921, 7th Cir. (1996).)

The first thing to note in this directive is that it is not a population-based criterion, but one that appeals to the individual case of the plaintiff. The second thing to note is that it is formulated in counterfactual terminology, using idioms such as "would have taken," "had the employee been," and "had been the same." What do they mean? Can one ever prove how an employer would have acted had Mary been straight? Certainly, this is not a variable that we can intervene upon in an experimental setting. Can data from an observational study prove an employer discriminating?

It turns out that Mary's case, though superficially different from Example 4.4.3, has a lot in common with the problem Mrs Smith faced over her unsuccessful cancer treatment. The probability that Mary's nonhiring is due to her sexual orientation can, similarly to Mrs Smith's cancer treatment, be expressed using the probability of sufficiency:

$$PS = P(Y_1 = 1|X = 0, Y = 0)$$

In this case, Y stands for Mary's hiring, and X stands for the interviewer's perception of Mary's sexual orientation. The expression reads: "the probability that Mary would have been hired had the interviewer perceived her as straight, given that the interviewer perceived her as gay, and she was not hired." (Note that the variable in question is the interviewer's *perception* of Mary's sexual orientation, not the orientation itself, because an intervention on perception is quite simple in this case—we need only to imagine that Mary never mentioned that she is gay; hypothesizing a change in Mary's actual orientation, although formally acceptable, brings with it an aura of awkwardness.)

We show in 4.5.2 that, although discrimination cannot be proved in individual cases, the probability that such discrimination took place can be determined, and this probability may sometimes reach a level approaching certitude. The next example examines how the problem of discrimination—in this case on gender, not sexual orientation may appear to a policy maker, rather than a juror.

4.4.5 Mediation and Path-disabling Interventions

Example 4.4.5 *A policy maker wishes to assess the extent to which gender disparity in hiring can be reduced by making hiring decisions gender-blind, rather than eliminating gender inequality in education or job training. The former concerns the "direct effect" of gender on hiring, whereas the latter concerns the "indirect effect," or the effect* mediated *via job qualification.*

In this example, fighting employers' prejudices and launching educational reforms are two contending policy options that involve costly investments and different implementation strategies. Knowing in advance which of the two, if successful, would have a greater impact on reducing hiring disparity is essential for planning, and depends critically on mediation analysis for resolution. For example, knowing that current hiring disparities are due primarily to employers' prejudices would render educational reforms superfluous, a fact that may save substantial resources. Note, however, that the policy decisions in this example concern the enabling and disabling of processes rather than lowering or raising values of specific variables. The educational reform program calls for disabling current educational practices and replacing them with a new program in which women obtain the same educational opportunities as men. The hiring-based proposal calls for disabling the current hiring process and replacing it with one in which gender plays no role in hiring decisions.

Because we are dealing with disabling processes rather than changing levels of variables, there is no way we can express the effect of such interventions using a *do*-operator, as we did in the mediation analysis of Section 3.7. We can express it, however, in a counterfactual language, using the desired end result as an antecedent. For example, if we wish to assess the hiring disparity after successfully implementing gender-blind hiring procedures, we impose the condition that all female applicants be treated like males as an antecedent and proceed to estimate the hiring rate under such a counterfactual condition.

The analysis proceeds as follows: the hiring status (Y) of a female applicant with qualification $Q = q$, given that the employer treats her as though she is a male is captured by the counterfactual $Y_{X=1,Q=q}$, where $X = 1$ refers to being a male. But since the value q would vary among applicants, we need to average this quantity according to the distribution of female qualification, giving $\sum_q E[Y_{X=1,Q=q}]P(Q = q|X = 0)$. Male applicants would have a similar chance at hiring except that the average is governed by the distribution of male qualification, giving

$$\sum_q E[Y_{X=1,Q=q}]P(Q = q|X = 1)$$

If we subtract the two quantities, we get

$$\sum_q E[Y_{X=1,Q=q}][P(Q = q|X = 0) - P(Q = q|X = 1)]$$

which is the indirect effect of gender on hiring, mediated by qualification. We call this effect the natural indirect effect (NIE), because we allow the qualification Q to vary naturally from applicant to applicant, as opposed to the controlled direct effect in Chapter 3, where we held the mediator at a constant level for the entire population. Here we merely disable the capacity of Y to respond to X but leave its response to Q unaltered.

The next question to ask is whether such a counterfactual expression can be identified from data. It can be shown (Pearl 2001) that, in the absence of confounding the NIE can be estimated by conditional probabilities, giving

$$NIE = \sum_q E[Y|X = 1, Q = q][P(Q = q|X = 0) - P(Q = q|X = 1)]$$

This expression is known as the *mediation formula*. It measures the extent to which the effect of X on Y is *explained* by its effect on the mediator Q. Counterfactual analysis permits us to define and assess NIE by "freezing" the direct effect of X on Y, and allowing the mediator (Q) of each unit to react to X in a natural way, as if no freezing took place.

The mathematical tools necessary for estimating the various nuances of mediation are summarized in Section 4.5.

4.5 Mathematical Tool Kits for Attribution and Mediation

As we examined the practical applications of counterfactual analysis in Section 4.4, we noted several recurring patterns that shared mathematical expressions as well as methods of solution. The first was the effect of treatment on the treated, *ETT*, whose syntactic signature was the counterfactual expression $E[Y_x|X = x']$, with x and x' two distinct values of X. We showed that problems as varied as recruitment to a program (Section 4.4.1) and additive interventions (Example 4.4.2) rely on the estimation of this expression, and we have listed conditions under which estimation is feasible, as well as the resulting estimand (Eqs. (4.21) and (4.8)).

Another recurring pattern appeared in problems of attribution, such as personal decision problems (Example 4.4.3) and possible cases of discrimination (Example 4.4.4). Here, the pattern was the expression for the probability of necessity:

$$PN = P(Y_0 = 0|X = 1, Y = 1)$$

The probability of necessity also pops up in problems of legal liability, where it reads: "The probability that the damage would not have occurred had the action not been taken ($Y_0 = 0$), given that, in fact, the damage did occur ($Y = 1$) and the action was taken ($X = 1$)." Section 4.5.1 summarizes mathematical results that will enable readers to estimate (or bound) PN using a combination of observational and experimental data.

Finally, in questions of mediation (Example 4.4.5) the key counterfactual expression was

$$E[Y_{x,M_{x'}}]$$

which reads, "The expected outcome (Y) had the treatment been $X = x$ and, simultaneously, had the mediator M attained the value ($M_{x'}$) it would have attained had X been x'". Section 4.5.2 will list the conditions under which this "nested" counterfactual expression can be estimated, as well as the resulting estimands and their interpretations.

4.5.1 A Tool Kit for Attribution and Probabilities of Causation

Assuming binary events, with $X = x$ and $Y = y$ representing treatment and outcome, respectively, and $X = x'$, $Y = y'$ their negations, our target quantity is defined by the English sentence:

> "Find the probability that if X had been x', Y would be y', given that, in reality, X is x and Y is y."

Mathematically, this reads

$$PN(x,y) = P(Y_{x'} = y' | X = x, Y = y) \tag{4.27}$$

This counterfactual quantity, named "probability of necessity" (PN), captures the legal criterion of "but for," according to which judgment in favor of a plaintiff should be made if and only if it is "more probable than not" that the damage would not have occurred *but for* the defendant's action (Robertson 1997).

Having written a formal expression for PN, Eq. (4.27), we can move on to the identification phase and ask what assumptions permit us to identify PN from empirical studies, be they observational, experimental, or a combination thereof.

Mathematical analysis of this problem (described in (Pearl 2000, Chapter 9)) yields the following results:

Theorem 4.5.1 *If Y is monotonic relative to X, that is, $Y_1(u) \geq Y_0(u)$ for all u, then PN is identifiable whenever the causal effect $P(y|do(x))$ is identifiable, and*

$$PN = \frac{P(y) - P(y|do(x'))}{P(x,y)} \tag{4.28}$$

or, substituting $P(y) = P(y|x)P(x) + P(y|x')(1 - P(x))$, we obtain

$$PN = \frac{P(y|x) - P(y|x')}{P(y|x)} + \frac{P(y|x') - P(y|do(x'))}{P(x,y)} \tag{4.29}$$

The first term on the r.h.s. of (4.29) is called the *excess risk ratio* (ERR) and is often used in court cases in the absence of experimental data (Greenland 1999). It is also known as the Attributable Risk Fraction among the exposed (Jewell 2004, Chapter 4.7). The second term (the *confounding factor* (CF)) represents a *correction* needed to account for confounding bias, that is, $P(y|do(x')) \neq P(y|x')$. Put in words, confounding occurs when the proportion of population for whom $Y = y$, when X is set to x' for everyone is not the same as the proportion of the population for whom $Y = y$ among those acquiring $X = x'$ by choice. For instance, suppose there is a case brought against a car manufacturer, claiming that its car's faulty design led to a man's death in a car crash. The ERR tells us how much more likely people are to die in crashes when driving one of the manufacturer's cars. If it turns out that people who buy the manufacturer's cars are more likely to drive fast (leading to deadlier crashes) than the general population, the second term will correct for that bias.

Equation (4.29) thus provides an estimable measure of necessary causation, which can be used for monotonic $Y_x(u)$ whenever the causal effect $P(y|do(x))$ can be estimated, be it from randomized trials or from graph-assisted observational studies (e.g., through the backdoor criterion). More significantly, it has also been shown (Tian and Pearl 2000) that the expression in (4.28) provides a lower bound for PN in the general nonmonotonic case. In particular, the upper and lower bounds on PN are given by

$$\max\left\{0, \frac{P(y) - P(y|do(x'))}{P(x,y)}\right\} \leq PN \leq \min\left\{1, \frac{P(y'|do(x')) - P(x',y')}{P(x,y)}\right\} \tag{4.30}$$

In drug-related litigation, it is not uncommon to obtain data from both experimental and observational studies. The former is usually available from the manufacturer or the agency that approved the drug for distribution (e.g., FDA), whereas the latter is often available from surveys of the population.

A few algebraic steps allow us to express the lower bound (*LB*) and upper bound (*UB*) as

$$LB = ERR + CF$$
$$UB = ERR + q + CF \tag{4.31}$$

where *ERR*, *CF*, and q are defined as follows:

$$CF \triangleq [P(y|x') - P(y_{x'})]/P(x, y) \tag{4.32}$$
$$ERR \triangleq 1 - 1/RR = 1 - P(y|x')/P(y|x) \tag{4.33}$$
$$q \triangleq P(y'|x)/P(y|x) \tag{4.34}$$

Here, CF represents the normalized degree of confounding among the unexposed ($X = x'$), ERR is the "excess risk ratio" and q is the ratio of negative to positive outcomes among the exposed.

Figure 4.5(a) and (b) depicts these bounds as a function of ERR, and reveals three useful features. First, regardless of confounding, the interval *UB–LB* remains constant and depends on only one observable parameter, $P(y'|x)/P(y|x)$. Second, the CF may raise the lower bound to meet the criterion of "more probable than not," $PN > \frac{1}{2}$, when the ERR alone would not suffice. Lastly, the amount of "rise" to both bounds is given by *CF*, which is the only estimate needed from the experimental data; the causal effect $P(y_x) - P(y_{x'})$ is not needed.

Theorem 4.5.1 further assures us that, if monotonicity can be assumed, the upper and lower bounds coincide, and the gap collapses entirely, as shown in Figure 4.5(b). This collapse does not reflect $q = 0$, but a shift from the bounds of (4.30) to the identified conditions of (4.28).

If it is the case that the experimental and survey data have been drawn at random from the same population, then the experimental data can be used to estimate the counterfactuals

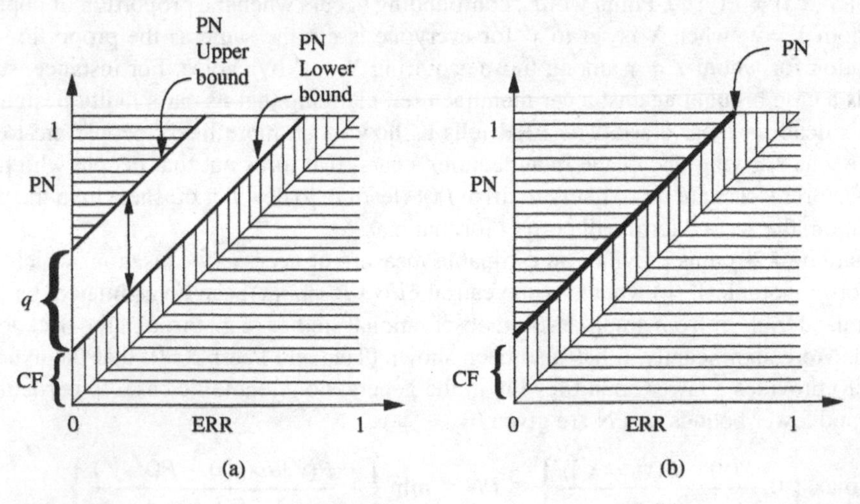

Figure 4.5 (a) Showing how probabilities of necessity (PN) are bounded, as a function of the excess risk ratio (ERR) and the confounding factor (CF) (Eq. (4.31)); (b) showing how PN is identified when monotonicity is assumed (Theorem 4.5.1)

of interest, for example, $P(Y_x = y)$, for the observational as well as experimental sampled populations.

Example 4.5.1 (Attribution in Legal Setting) *A lawsuit is filed against the manufacturer of drug x, charging that the drug is likely to have caused the death of Mr A, who took it to relieve back pains. The manufacturer claims that experimental data on patients with back pains show conclusively that drug x has only minor effects on death rates. However, the plaintiff argues that the experimental study is of little relevance to this case because it represents average effects on patients in the study, not on patients like Mr A who did not participate in the study. In particular, argues the plaintiff, Mr A is unique in that he used the drug of his own volition, unlike subjects in the experimental study, who took the drug to comply with experimental protocols. To support this argument, the plaintiff furnishes nonexperimental data on patients who, like Mr A, chose drug x to relieve back pains but were not part of any experiment, and who experienced lower death rates than those who didn't take the drug. The court must now decide, based on both the experimental and nonexperimental studies, whether it is "more probable than not" that drug x was in fact the cause of Mr A's death.*

To illustrate the usefulness of the bounds in Eq. (4.30), consider (hypothetical) data associated with the two studies shown in Table 4.5. (In the analyses below, we ignore sampling variability.)

The experimental data provide the estimates

$$P(y|do(x)) = 16/1000 = 0.016 \tag{4.35}$$

$$P(y|do(x')) = 14/1000 = 0.014 \tag{4.36}$$

whereas the nonexperimental data provide the estimates

$$P(y) = 30/2000 = 0.015 \tag{4.37}$$

$$P(x, y) = 2/2000 = 0.001 \tag{4.38}$$

$$P(y|x) = 2/1000 = 0.002 \tag{4.39}$$

$$P(y|x') = 28/1000 = 0.028 \tag{4.40}$$

Table 4.5 Experimental and nonexperimental data used to illustrate the estimation of PN, the probability that drug x was responsible for a person's death (y)

	Experimental		Nonexperimental	
	$do(x)$	$do(x')$	x	x'
Deaths (y)	16	14	2	28
Survivals (y')	984	986	998	972

Assuming that drug x can only cause (but never prevent) death, monotonicity holds, and Theorem 4.5.1 (Eq. (4.29)) yields

$$PN = \frac{P(y|x) - P(y|x')}{P(y|x)} + \frac{P(y|x') - P(y|do(x'))}{P(x,y)}$$

$$= \frac{0.002 - 0.028}{0.002} + \frac{0.028 - 0.014}{0.001} = -13 + 14 = 1 \tag{4.41}$$

We see that while the observational ERR is negative (−13), giving the impression that the drug is actually preventing deaths, the bias-correction term (+14) rectifies this impression and sets the probability of necessity (PN) to unity. Moreover, since the lower bound of Eq. (4.30) becomes 1, we conclude that $PN = 1.00$ even without assuming monotonicity. Thus, the plaintiff was correct; barring sampling errors, the data provide us with 100% assurance that drug x was in fact responsible for the death of Mr A.

To complete this tool kit for attribution, we note that the other two probabilities that came up in the discussion on personal decision-making (Example 4.4.3), PS and PNS, can be bounded by similar expressions; see (Pearl 2000, Chapter 9) and (Tian and Pearl 2000).

In particular, when $Y_x(u)$ is monotonic, we have

$$PNS = P(Y_x = 1, Y_{x'} = 0)$$

$$= P(Y_x = 1) - P(Y_{x'} = 1) \tag{4.42}$$

as asserted in Example 4.4.3, Eq. (4.26).

Study questions

Study question 4.5.1

Consider the dilemma faced by Ms Jones, as described in Example 4.4.3. Assume that, in addition to the experimental results of Fisher et al. (2002), she also gains access to an observational study, according to which the probability of recurrent tumor in all patients (regardless of irradiation) is 30%, whereas among the recurrent cases, 70% did not choose therapy. Use the bounds provided in Eq. (4.30) to update her estimate that her decision was necessary for remission.

4.5.2 A Tool Kit for Mediation

The canonical model for a typical mediation problem takes the form:

$$t = f_T(u_T) \quad m = f_M(t, u_M) \quad y = f_Y(t, m, u_Y) \tag{4.43}$$

where T (treatment), M (mediator), and Y (outcome) are discrete or continuous random variables, f_T, f_M, and f_Y are arbitrary functions, and U_T, U_M, U_Y represent, respectively, omitted factors that influence T, M, and Y. The triplet $U = (U_T, U_M, U_Y)$ is a random vector that accounts for all variations among individuals.

In Figure 4.6(a), the omitted factors are assumed to be arbitrarily distributed but mutually independent. In Figure 4.6(b), the dashed arcs connecting U_T and U_M (as well as U_M and U_Y) encode the understanding that the factors in question may be dependent.

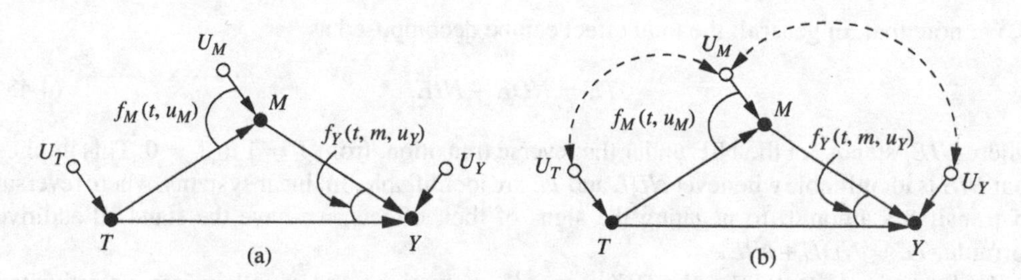

Figure 4.6 (a) The basic nonparametric mediation model, with no confounding. (b) A confounded mediation model in which dependence exists between U_M and (U_T, U_Y)

Counterfactual definition of direct and indirect effects

Using the structural model of Eq. (4.43) and the counterfactual notation defined in Section 4.2.1, four types of effects can be defined for the transition from $T = 0$ to $T = 1$. Generalizations to arbitrary reference points, say from $T = t$ to $T = t'$, are straightforward[1]:

(a) Total effect –

$$\begin{aligned} TE &= E[Y_1 - Y_0] \\ &= E[Y|do(T = 1)] - E[Y|do(T = 0)] \end{aligned} \tag{4.44}$$

TE measures the expected increase in Y as the treatment changes from $T = 0$ to $T = 1$, while the mediator is allowed to track the change in T naturally, as dictated by the function f_M.

(b) Controlled direct effect –

$$\begin{aligned} CDE(m) &= E[Y_{1,m} - Y_{0,m}] \\ &= E[Y|do(T = 1, M = m)] - E[Y|do(T = 0, M = m)] \end{aligned} \tag{4.45}$$

CDE measures the expected increase in Y as the treatment changes from $T = 0$ to $T = 1$, while the mediator is set to a specified level $M = m$ uniformly over the entire population.

(c) Natural direct effect –

$$NDE = E[Y_{1,M_0} - Y_{0,M_0}] \tag{4.46}$$

NDE measures the expected increase in Y as the treatment changes from $T = 0$ to $T = 1$, while the mediator is set to whatever value it *would have attained* (for each individual) prior to the change, that is, under $T = 0$.

(d) Natural indirect effect –

$$NIE = E[Y_{0,M_1} - Y_{0,M_0}] \tag{4.47}$$

NIE measures the expected increase in Y when the treatment is held constant, at $T = 0$, and M changes to whatever value it would have attained (for each individual) under $T = 1$. It captures, therefore, the portion of the effect that can be explained by mediation alone, while disabling the capacity of Y to respond to T.

[1] These definitions apply at the population levels; the unit-level effects are given by the expressions under the expectation. All expectations are taken over the factors U_M and U_Y.

We note that, in general, the total effect can be decomposed as

$$TE = NDE - NIE_r \tag{4.48}$$

where NIE_r stands for the NIE under the reverse transition, from $T = 1$ to $T = 0$. This implies that *NIE* is identifiable whenever *NDE* and *TE* are identifiable. In linear systems, where reversal of transitions amounts to negating the signs of their effects, we have the standard additive formula, $TE = NDE + NIE$.

We further note that *TE* and $CDE(m)$ are *do*-expressions and can, therefore, be estimated from experimental data or in observational studies using the backdoor or front-door adjustments. Not so for the *NDE* and *NIE*; a new set of assumptions is needed for their identification.

Conditions for identifying natural effects

The following set of conditions, marked *A*-1 to *A*-4, are sufficient for identifying both direct and indirect natural effects.

We can identify the *NDE* and *NIE* provided that there exists a set *W* of measured covariates such that

A-1 No member of *W* is a descendant of *T*.
A-2 *W* blocks all backdoor paths from *M* to *Y* (after removing $T \to M$ and $T \to Y$).
A-3 The *W*-specific effect of *T* on *M* is identifiable (possibly using experiments or adjustments).
A-4 The *W*-specific joint effect of $\{T, M\}$ on *Y* is identifiable (possibly using experiments or adjustments).

Theorem 4.5.2 (Identification of the *NDE*) *When conditions A-1 and A-2 hold, the natural direct effect is experimentally identifiable and is given by*

$$\begin{aligned} NDE = &\sum_m \sum_w [E[Y|do(T = 1, M = m), W = w] - E[Y|do(T = 0, M = m), W = w]] \\ &\times P(M = m|do(T = 0), W = w)P(W = w) \end{aligned} \tag{4.49}$$

The identifiability of the do-expressions in Eq. (4.49) is guaranteed by conditions A-3 and A-4 and can be determined using the backdoor or front-door criteria.

Corollary 4.5.1 *If conditions A-1 and A-2 are satisfied by a set W that also deconfounds the relationships in A-3 and A-4, then the do-expressions in Eq. (4.49) are reducible to conditional expectations, and the natural direct effect becomes*

$$\begin{aligned} NDE = &\sum_m \sum_w [E[Y|T = 1, M = m, W = w] - E[Y|T = 0, M = m, W = w]] \\ &\times P(M = m|T = 0, W = w)P(W = w) \end{aligned} \tag{4.50}$$

In the nonconfounding case (Figure 4.6(a)), *NDE* reduces to

$$NDE = \sum_m [E[Y \mid T = 1, M = m] - E[Y \mid T = 0, M = m]]P(M = m \mid T = 0). \tag{4.51}$$

Similarly, using (4.48) and $TE = E[Y|T=1] - E[Y|T=0]$, NIE becomes

$$NIE = \sum_{m} E[Y \mid T=0, M=m][P(M=m \mid T=1) - P(M=m \mid T=0)] \tag{4.52}$$

The last two expressions are known as the *mediation formulas*. We see that while NDE is a weighted average of CDE, no such interpretation can be given to NIE.

The counterfactual definitions of NDE and NIE (Eqs. (4.46) and (4.47)) permit us to give these effects meaningful interpretations in terms of "response fractions." The ratio NDE/TE measures the fraction of the response that is transmitted directly, with M "frozen." NIE/TE measures the fraction of the response that may be transmitted through M, with Y blinded to T. Consequently, the difference $(TE - NDE)/TE$ measures the fraction of the response that is necessarily due to M.

Numerical example: Mediation with binary variables

To anchor these mediation formulas in a concrete example, we return to the encouragement-design example of Section 4.2.3 and assume that $T = 1$ stands for participation in an enhanced training program, $Y = 1$ for passing the exam, and $M = 1$ for a student spending more than 3 hours per week on homework. Assume further that the data described in Tables 4.6 and 4.7 were obtained in a randomized trial with no mediator-to-outcome confounding (Figure 4.6(a)). The data shows that training tends to increase both the time spent on homework and the rate of success on the exam. Moreover, training and time spent on homework together are more likely to produce success than each factor alone.

Our research question asks for the extent to which students' homework contributes to their increased success rates regardless of the training program. The policy implications of such questions lie in evaluating policy options that either curtail or enhance homework efforts, for example, by counting homework effort in the final grade or by providing students with

Table 4.6 The expected success (Y) for treated ($T = 1$) and untreated ($T = 0$) students, as a function of their homework (M)

Treatment T	Homework M	Success rate $E(Y \mid T=t, M=m)$
1	1	0.80
1	0	0.40
0	1	0.30
0	0	0.20

Table 4.7 The expected homework (M) done by treated ($T = 1$) and untreated ($T = 0$) students

Treatment T	Homework $E(M \mid T=t)$
0	0.40
1	0.75

adequate work environments at home. An extreme explanation of the data, with significant impact on educational policy, might argue that the program does not contribute substantively to students' success, save for encouraging students to spend more time on homework, an encouragement that could be obtained through less expensive means. Opposing this theory, we may have teachers who argue that the program's success is substantive, achieved mainly due to the unique features of the curriculum covered, whereas the increase in homework efforts cannot alone account for the success observed.

Substituting the data into Eqs. (4.51) and (4.52) gives

$$NDE = (0.40 - 0.20)(1 - 0.40) + (0.80 - 0.30)0.40 = 0.32$$

$$NIE = (0.75 - 0.40)(0.30 - 0.20) = 0.035$$

$$TE = 0.80 \times 0.75 + 0.40 \times 0.25 - (0.30 \times 0.40 + 0.20 \times 0.60) = 0.46$$

$$NIE/TE = 0.07, NDE/TE = 0.696, 1 - NDE/TE = 0.304$$

We conclude that the program as a whole has increased the success rate by 46% and that a significant portion, 30.4%, of this increase is due to the capacity of the program to stimulate improved homework effort. At the same time, only 7% of the increase can be explained by stimulated homework alone without the benefit of the program itself.

Study questions

Study question 4.5.2

Consider the structural model:

$$y = \beta_1 m + \beta_2 t + u_y \quad (4.53)$$

$$m = \gamma_1 t + u_m \quad (4.54)$$

(a) Use the basic definition of the natural effects (Eqs. (4.46) and (4.47)) to determine TE, NDE, and NIE.

(b) Repeat (a) assuming that u_y is correlated with u_m.

Study question 4.5.3

Consider the structural model:

$$y = \beta_1 m + \beta_2 t + \beta_3 tm + \beta_4 w + u_y \quad (4.55)$$

$$m = \gamma_1 t + \gamma_2 w + u_m \quad (4.56)$$

$$w = \alpha t + u_w \quad (4.57)$$

with $\beta_3 tm$ representing an interaction term.

(a) Use the basic definition of the natural effects (Eqs. (4.46) and (4.47)) (treating M as the mediator), to determine the portion of the effect for which mediation is necessary *(TE − NDE) and the portion for which mediation is* sufficient *(NIE). Hint: Show that:*

$$NDE = \beta_2 + \alpha\beta_4 \tag{4.58}$$

$$NIE = \beta_1(\gamma_1 + \alpha\gamma_2) \tag{4.59}$$

$$TE = \beta_2 + (\gamma_1 + \alpha\gamma_2)(\beta_3 + \beta_1) + \alpha\beta_4 \tag{4.60}$$

$$TE - NDE = (\beta_1 + \beta_3)(\gamma_1 + \alpha\gamma_2) \tag{4.61}$$

(b) Repeat, using W as the mediator.

Study question 4.5.4

Apply the mediation formulas provided in this section to the discrimination case discussed in Section 4.4.4, and determine the extent to which ABC International practiced discrimination in their hiring criteria. Use the data in Tables 4.6 and 4.7, with $T = 1$ standing for male applicants, $M = 1$ standing for highly qualified applicants, and $Y = 1$ standing for hiring. (Find the proportion of the hiring disparity that is due to gender, and the proportion that could be explained by disparity in qualification alone.)

Ending Remarks

The analysis of mediation is perhaps the best arena to illustrate the effectiveness of the counterfactual-graphical symbiosis that we have been pursuing in this book. If we examine the identifying conditions *A*-1 to *A*-4, we find four assertions about the model that are not too easily comprehended. To judge their plausibility in any given scenario, without the graph before us, is unquestionably a formidable, superhuman task. Yet the symbiotic analysis frees investigators from the need to understand, articulate, examine, and judge the plausibility of the assumptions needed for identification. Instead, the method can confirm or disconfirm these assumptions algorithmically from a more reliable set of assumptions, those encoded in the structural model itself. Once constructed, the causal diagram allows simple path-tracing routines to replace much of the human judgment deemed necessary in mediation analysis; the judgment invoked in the construction of the diagrams is sufficient, and that construction requires only judgment about causal relationships among realizable variables and their disturbances.

Bibliographical Notes for Chapter 4

The definition of counterfactuals as derivatives of structural equations, Eq. (4.5), was introduced by Balke and Pearl (1994a,b), who applied it to the estimation of probabilities of causation in legal settings. The philosopher David Lewis defined counterfactuals in terms of similarity among possible worlds Lewis (1973). In statistics, the notation $Y_x(u)$ was devised by Neyman (1923), to denote the potential response of unit u in a controlled randomized trial, under treatment $X = x$. It remained relatively unnoticed until Rubin (1974) treated Y_x as a random variable and connected it to observed variable via the consistency rule of Eq. (4.6), which is a theorem in both Lewis's logic and in structural models. The relationships among these three formalisms of counterfactuals are discussed at length in Pearl (2000, Chapter 7), where they are shown to be logically equivalent; a problem solved in one framework would yield the same solution in another. Rubin's framework, known as "potential outcomes," differs from the structural account only in the language in which problems are defined, hence, in the mathematical tools available for their solution. In the potential outcome framework, problems are defined algebraically as assumptions about counterfactual independencies, also known as "ignorability assumptions." These types of assumptions, exemplified in Eq. (4.15), may become too complicated to interpret or verify by unaided judgment. In the structural framework, on the other hand, problems are defined in the form of causal graphs, from which dependencies of counterfactuals (e.g., Eq. (4.15)) can be derived mechanically. The reason some statisticians prefer the algebraic approach is, primarily, because graphs are relatively new to statistics. Recent books in social science (e.g., Morgan and Winship 2014) and in health science (e.g., VanderWeele 2015) are taking the hybrid, graph-counterfactual approach pursued in our book.

The section on linear counterfactuals is based on Pearl (2009, pp. 389–391). Recent advances are provided in Cai and Kuroki (2006) and Chen and Pearl (2014). Our discussion of ETT (Effect of Treatment on the Treated), as well as additive interventions, is based on Shpitser and Pearl (2009), which provides a full characterization of models in which ETT is identifiable.

Legal questions of attribution, as well as probabilities of causation are discussed at length in Greenland (1999) who pioneered the counterfactual approach to such questions. Our treatment of *PN*, *PS*, and *PNS* is based on Tian and Pearl (2000) and Pearl (2000, Chapter 9). Recent results, including the tool kit of Section 4.5.1, are given in Pearl (2015a).

Mediation analysis (Sections 4.4.5 and 4.5.2), as we remarked in Chapter 3, has a long tradition in the social sciences (Duncan 1975; Kenny 1979), but has gone through a dramatic revolution through the introduction of counterfactual analysis. A historical account of the conceptual transition from the statistical approach of Baron and Kenny (1986) to the modern, counterfactual-based approach of natural direct and indirect effects (Pearl 2001; Robins and Greenland 1992) is given in Sections 1 and 2 of Pearl (2014a). The recent text of VanderWeele (2015) enhances this development with new results and new applications. Additional advances in mediation, including sensitivity analysis, bounds, multiple mediators, and stronger identifying assumptions are discussed in Imai et al. (2010) and Muthén and Asparouhov (2015).

The mediation tool kit of Section 4.5.2 is based on Pearl (2014a). Shpitser (2013) has derived a general criterion for identifying indirect effects in graphs.

References

Balke A and Pearl J 1994a Counterfactual probabilities: Computational methods, bounds, and applications. In *Uncertainty in Artificial Intelligence 10* (ed. de Mantaras RL and Poole D) Morgan Kaufmann Publishers, San Mateo, CA pp. 46–54.

Balke A and Pearl J 1994b Probabilistic evaluation of counterfactual queries. *Proceedings of the Twelfth National Conference on Artificial Intelligence*, vol. **I**, MIT Press, Menlo Park, CA pp. 230–237.

Bareinboim E and Pearl J 2012 Causal inference by surrogate experiments (or, *z*-identifiability). *Proceedings of the Twenty-eighth Conference on Uncertainty in Artificial Intelligence* (ed. de Freitas N and Murphy K) AUAI Press, Corvallis, OR, pp. 113–120.

Bareinboim E and Pearl J 2013 A general algorithm for deciding transportability of experimental results. *Journal of Causal Inference* **1** (1), 107–134.

Bareinboim E and Pearl J 2016 Causal inference and the data-fusion problem. *Proceedings of the National Academy of Sciences* **113** (17), 7345–7352.

Bareinboim E, Tian J and Pearl J 2014 Recovering from selection bias in causal and statistical inference. *Proceedings of the Twenty-eighth AAAI Conference on Artificial Intelligence* (ed. Brodley CE and Stone P) AAAI Press, Palo Alto, CA, pp. 2410–2416.

Baron R and Kenny D 1986 The moderator-mediator variable distinction in social psychological research: Conceptual, strategic, and statistical considerations. *Journal of Personality and Social Psychology* **51** (6), 1173–1182.

Berkson J 1946 Limitations of the application of fourfold table analysis to hospital data. *Biometrics Bulletin* **2**, 47–53.

Bollen K 1989 *Structural Equations with Latent Variables*. John Wiley & Sons, Inc., New York.

Bollen K and Pearl J 2013 Eight myths about causality and structural equation models. In *Handbook of Causal Analysis for Social Research* (ed. Morgan S) Springer-Verlag, Dordrecht, Netherlands pp. 245–274.

Bowden R and Turkington D 1984 *Instrumental Variables*. Cambridge University Press, Cambridge, England.

Brito C and Pearl J 2002 Generalized instrumental variables. *Uncertainty in Artificial Intelligence, Proceedings of the Eighteenth Conference* (ed. Darwiche A and Friedman N) Morgan Kaufmann San Francisco, CA pp. 85–93.

Cai Z and Kuroki M 2006 Variance estimators for three 'probabilities of causation'. *Risk Analysis* **25** (6), 1611–1620.

Causal Inference in Statistics: A Primer, First Edition. Judea Pearl, Madelyn Glymour, and Nicholas P. Jewell.

Companion Website: www.wiley.com/go/Pearl/Causality

Chen B and Pearl J 2014 *Graphical tools for linear structural equation modeling*. Technical Report R-432, Department of Computer Science, University of California, Los Angeles, CA. Submitted, Psychometrika, http://ftp.cs.ucla.edu/pub/stat_ser/r432.pdf.

Cole S and Hernán M 2002 Fallibility in estimating direct effects. *International Journal of Epidemiology* **31** (1), 163–165.

Conrady S and Jouffe L 2015 *Bayesian Networks and BayesiaLab: A Practical Introduction for Researchers* 1st edition edn. Bayesia USA.

Cox D 1958 *The Planning of Experiments*. John Wiley and Sons, New York.

Darwiche A 2009 *Modeling and Reasoning with Bayesian Networks*. Cambridge University Press, New York.

Duncan O 1975 *Introduction to Structural Equation Models*. Academic Press, New York.

Elwert F 2013 Graphical causal models. In *Handbook of Causal Analysis for Social Research* (ed. Morgan S) Springer-Verlag, Dordrecht, Netherlands pp. 245–274.

Fenton N and Neil M 2013 *Risk Assessment and Decision Analysis with Bayesian Networks*. CRC Press, Boca Raton, FL.

Fisher R 1922 On the mathematical foundations of theoretical statistics. *Philosophical Transactions of the Royal Society of London, Series A* **222**, 311.

Fisher B, Anderson S, Bryant J, Margolese RG, Deutsch M, Fisher ER, Jeong JH and Wolmark N 2002 Twenty-year follow-up of a randomized trial comparing total mastectomy, lumpectomy, and lumpectomy plus irradiation for the treatment of invasive breast cancer. *New England Journal of Medicine* **347** (16), 1233–1241.

Glymour MM 2006 Using causal diagrams to understand common problems in social epidemiology. *Methods in Social Epidemiology* John Wiley & Sons, Inc., San Francisco, CA pp. 393–428.

Glymour M and Greenland S 2008 Causal diagrams. In *Modern Epidemiology* (ed. Rothman K, Greenland S, and Lash T) 3rd edn. Lippincott Williams & Wilkins Philadelphia, PA pp. 183–209.

Greenland S 1999 Relation of probability of causation, relative risk, and doubling dose: A methodologic error that has become a social problem. *American Journal of Public Health* **89** (8), 1166–1169.

Greenland S 2000 An introduction to instrumental variables for epidemiologists. *International Journal of Epidemiology* **29** (4), 722–729.

Grinstead CM and Snell JL 1998 *Introduction to Probability* second revised edn. American Mathematical Society, United States.

Haavelmo T 1943 The statistical implications of a system of simultaneous equations. *Econometrica* **11**, 1–12. Reprinted in DF Hendry and MS Morgan (Eds.), 1995 *The Foundations of Econometric Analysis*, Cambridge University Press pp. 477–490.

Hayduk L, Cummings G, Stratkotter R, Nimmo M, Grygoryev K, Dosman D, Gilespie, M., Pazderka-Robinson H and Boadu K 2003 Pearl's *d*-separation: One more step into causal thinking. *Structural Equation Modeling* **10** (2), 289–311.

Heise D 1975 *Causal Analysis*. John Wiley and Sons, New York.

Hernán M and Robins J 2006 Estimating causal effects from epidemiological data. *Journal of Epidemiology and Community Health* **60** (7), 578–586. DOI: 10.1136/jech.2004.029496.

Hernández-Díaz S, Schisterman E and Hernán M 2006 The birth weight "paradox" uncovered? *American Journal of Epidemiology* **164** (11), 1115–1120.

Holland P 1986 Statistics and causal inference. *Journal of the American Statistical Association* **81** (396), 945–960.

Howard R and Matheson J 1981 Influence diagrams. In *Principles and Applications of Decision Analysis* (ed. Howard R and Matheson J) Strategic Decisions Group, Menlo Park, CA pp.721–762.

Imai K, Keele L and Yamamoto T 2010 Identification, inference, and sensitivity analysis for causal mediation effects. *Statistical Science* **25** (1), 51–71.

Jewell NP 2004 *Statistics for Epidemiology*. Chapman & Hall/CRC, Boca Raton, FL.

Kenny D 1979 *Correlation and Causality*. John Wiley & Sons, Inc., New York.

Kiiveri H, Speed T and Carlin J 1984 Recursive causal models. *Journal of Australian Math Society* **36**, 30–52.

Kim J and Pearl J 1983 A computational model for combined causal and diagnostic reasoning in inference systems. *Proceedings of the Eighth International Joint Conference on Artificial Intelligence (IJCAI-83)*, pp. 190–193, Karlsruhe, Germany.

Kline RB 2016 *Principles and Practice of Structural Equation Modeling* fourth: Revised and expanded edn. Guilford Publications, Inc., New York.

Koller K and Friedman N 2009 *Probabilistic Graphical Models: Principles and Techniques*. MIT Press, United States.

Kyono T 2010 *Commentator: A front-end user-interface module for graphical and structural equation modeling*. Master's thesis Department of Computer Science, University of California, Los Angeles, CA.

Lauritzen S 1996 *Graphical Models*. Clarendon Press, Oxford. Reprinted 2004 with corrections.

Lewis D 1973 Causation. *Journal of Philosophy* **70**, 556–567.

Lindley DV 2014 *Understanding Uncertainty* revised edn. John Wiley & Sons, Inc., Hoboken, NJ.

Lord FM 1967 A paradox in the interpretation of group comparisons. *Psychological Bulletin* **68**, 304–305.

Mohan K, Pearl J and Tian J 2013 Graphical models for inference with missing data. In *Advances in Neural Information Processing Systems 26* (ed. Burges C, Bottou L, Welling M, Ghahramani Z and Weinberger K) Neural Information Processing Systems Foundation, Inc. pp. 1277–1285.

Moore D, McCabe G and Craig B 2014 *Introduction to the Practice of Statistics*. W.H. Freeman & Co., New York.

Morgan SL and Winship C 2014 *Counterfactuals and Causal Inference: Methods and Principles for Social Research, Analytical Methods for Social Research* 2nd edn. Cambridge University Press, New York.

Muthén B and Asparouhov T 2015 Causal effects in mediation modeling: An introduction with applications to latent variables. *Structural Equation Modeling: A Multidisciplinary Journal* **22** (1), 12–23.

Neyman J 1923 On the application of probability theory to agricultural experiments. Essay on principles. Section 9. *Statistical Science* **5** (4), 465–480.

Pearl J 1985 Bayesian networks: A model of self-activated memory for evidential reasoning. *Proceedings, Cognitive Science Society*, pp. 329–334, Irvine, CA.

Pearl J 1986 Fusion, propagation, and structuring in belief networks. *Artificial Intelligence* **29**, 241–288.

Pearl J 1988 *Probabilistic Reasoning in Intelligent Systems*. Morgan Kaufmann, San Mateo, CA.

Pearl J 1993 Comment: Graphical models, causality, and intervention. *Statistical Science* **8** (3), 266–269.

Pearl J 1995 Causal diagrams for empirical research. *Biometrika* **82** (4), 669–710.

Pearl J 1998 Graphs, causality, and structural equation models. *Sociological Methods and Research* **27** (2), 226–284.

Pearl J 2000 *Causality: Models, Reasoning, and Inference*. Cambridge University Press, New York.

Pearl J 2001 Direct and indirect effects. *Proceedings of the Seventeenth Conference on Uncertainty in Artificial Intelligence* Morgan Kaufmann San Francisco, CA pp. 411–420.

Pearl J 2009 *Causality: Models, Reasoning, and Inference* 2nd edn. Cambridge University Press, New York.

Pearl J 2014a Interpretation and identification of causal mediation. *Psychological Methods* **19**, 459–481.

Pearl J 2014b Understanding Simpson's paradox. *The American Statistician* **88** (1), 8–13.

Pearl J 2015a Causes of effects and effects of causes. *Journal of Sociological Methods and Research* **44**, 149–164.

Pearl J 2015b Detecting latent heterogeneity. *Sociological Methods and Research* DOI: 10.1177/0049124115600597, online:1–20.

Pearl J 2015c Trygve Haavelmo and the emergence of causal calculus. *Econometric Theory*, Special issue on Haavelmo Centennial **31** (1), 152–179.

Pearl J 2016 *Lord's paradox revisited—(oh Lord! Kumbaya!). Journal of Causal Inference* **4** (2). DOI: 10.1515/jci-2016-0021.

Pearl J and Bareinboim E 2014 External validity: From *do*-calculus to transportability across populations. *Statistical Science* **29** ,579–595.

Pearl J and Mackenzie D 2018 *The Book of Why: The New Science of Cause and Effect*. Basic Books, New York.

Pearl J and Paz A 1987 GRAPHOIDS: A graph-based logic for reasoning about relevance relations. In *Advances in Artificial Intelligence-II* (ed. Duboulay B, Hogg D and Steels L) North-Holland Publishing Co. pp. 357–363.

Pearl J and Robins J 1995 Probabilistic evaluation of sequential plans from causal models with hidden variables. In *Uncertainty in Artificial Intelligence 11* (ed. Besnard P and Hanks S) Morgan Kaufmann, San Francisco, CA pp. 444–453.

Pearl J and Verma T 1991 A theory of inferred causation. *Principles of Knowledge Representation and Reasoning: Proceedings of the Second International Conference* (ed. Allena J, Fikes R and Sandewall E) Morgan Kaufmann San Mateo, CA pp. 441–452.

Pigou A 1911 *Alcoholism and Heredity*. Westminster Gazette. February 2.

Rebane G and Pearl J 1987 The recovery of causal poly-trees from statistical data. *Proceedings of the Third Workshop on Uncertainty in AI*, pp. 222–228, Seattle, WA.

Reichenbach H 1956 *The Direction of Time*. University of California Press, Berkeley, CA.

Robertson D 1997 The common sense of cause in fact. *Texas Law Review* **75** (7), 1765–1800.

Robins J 1986 A new approach to causal inference in mortality studies with a sustained exposure period—applications to control of the healthy workers survivor effect. *Mathematical Modeling* **7**, 1393–1512.

Robins J and Greenland S 1992 Identifiability and exchangeability for direct and indirect effects. *Epidemiology* **3** (2), 143–155.

Rubin D 1974 Estimating causal effects of treatments in randomized and nonrandomized studies. *Journal of Educational Psychology* **66**, 688–701.

Selvin S 2004 *Biostatistics: How it Works*. Pearson, New Jersey.

Senn S 2006 Change from baseline and analysis of covariance revisited. *Statistics in Medicine* **25**, 4334–4344.

Shpitser I 2013 Counterfactual graphical models for longitudinal mediation analysis with unobserved confounding. *Cognitive Science* **37** (6), 1011–1035.

Shpitser I and Pearl J 2007 What counterfactuals can be tested. *Proceedings of the Twenty-Third Conference on Uncertainty in Artificial Intelligence* AUAI Press Vancouver, BC, Canada pp. 352–359. Also, *Journal of Machine Learning Research* **9**, 1941–1979, 2008.

Shpitser I and Pearl J 2008 Complete identification methods for the causal hierarchy. *Journal of Machine Learning Research* **9**, 1941–1979.

Shpitser I and Pearl J 2009 Effects of treatment on the treated: Identification and generalization. *Proceedings of the Twenty-Fifth Conference on Uncertainty in Artificial Intelligence* AUAI Press Montreal, Quebec pp. 514–521.

Simon H 1953 Causal ordering and identifiability. In *Studies in Econometric Method* (ed. Hood WC and Koopmans T) John Wiley & Sons, Inc. New York pp. 49–74.

Simpson E 1951 The interpretation of interaction in contingency tables. *Journal of the Royal Statistical Society, Series B* **13**, 238–241.

Spirtes P and Glymour C 1991 An algorithm for fast recovery of sparse causal graphs. *Social Science Computer Review* **9** (1), 62–72.

Spirtes P, Glymour C and Scheines R 1993 *Causation, Prediction, and Search*. Springer-Verlag, New York.

Stigler SM 1999 *Statistics on the Table: The History of Statistical Concepts and Methods*. Harvard University Press, Cambridge, MA, Hoboken, NJ.

Strotz R and Wold H 1960 Recursive versus nonrecursive systems: An attempt at synthesis. *Econometrica* **28**, 417–427.

Textor J, Hardt J and Knuüppel S 2011 DAGitty: A graphical tool for analyzing causal diagrams. *Epidemiology* **22** (5), 745.

Tian J, Paz A and Pearl J 1998 *Finding minimal d-separators*. Technical Report R-254, Department of Computer Science, University of California, Los Angeles, CA. http://ftp.cs.ucla.edu/pub/stat_ser/r254 .pdf.

Tian J and Pearl J 2000 Probabilities of causation: bounds and identification. *Annals of Mathematics and Artificial Intelligence* **28**, 287–313.

Tian J and Pearl J 2002 A general identification condition for causal effects. *Proceedings of the Eighteenth National Conference on Artificial Intelligence* AAAI Press/The MIT Press Menlo Park, CA pp. 567–573.

VanderWeele T 2015 *Explanation in Causal Inference: Methods for Mediation and Interaction*. Oxford University Press, New York.

Verma T and Pearl J 1988 Causal networks: Semantics and expressiveness. *Proceedings of the Fourth Workshop on Uncertainty in Artificial Intelligence*, pp. 352–359, Mountain View, CA. Also in R. Shachter, T.S. Levitt, and L.N. Kanal (Eds.), *Uncertainty in AI 4*, Elsevier Science Publishers, 69–76, 1990.

Verma T and Pearl J 1990 Equivalence and synthesis of causal models. *Proceedings of the Sixth Conference on Uncertainty in Artificial Intelligence*, pp. 220–227, Cambridge, MA.

Virgil 29 BC Georgics. Verse 490, Book 2.

Wainer H 1991 Adjusting for differential base rates: Lord's paradox again. *Psychological Bulletin* **109**, 147–151.

Wooldridge J 2013 Introductory Econometrics: A Modern Approach 5th international edn. South-Western, Mason, OH.

Index

Causal Inference in Statistics: A Primer, First Edition. Judea Pearl, Madelyn Glymour, and Nicholas P. Jewell.

Companion Website: www.wiley.com/go/Pearl/Causality